W9-ADZ-747

Separation Methods in
Organic Chemistry and Biochemistry

Separation Methods in Organic Chemistry and Biochemistry

FRANK J. WOLF

Merck Sharp & Dohme Research Laboratories
Rahway, New Jersey

ACADEMIC PRESS New York and London 1969

COPYRIGHT © 1969, BY ACADEMIC PRESS, INC.
ALL RIGHTS RESERVED
NO PART OF THIS BOOK MAY BE REPRODUCED IN ANY FORM,
BY PHOTOSTAT, MICROFILM, RETRIEVAL SYSTEM, OR ANY
OTHER MEANS, WITHOUT WRITTEN PERMISSION FROM
THE PUBLISHERS.

ACADEMIC PRESS, INC.
111 Fifth Avenue, New York, New York 10003

United Kingdom Edition published by
ACADEMIC PRESS, INC. (LONDON) LTD.
Berkeley Square House, London W.1

LIBRARY OF CONGRESS CATALOG CARD NUMBER: 71-84256

PRINTED IN THE UNITED STATES OF AMERICA

QD
63
.S4
W6

Preface

The isolation of pure substance is frequently the major laboratory effort required for the solution of a chemical or biochemical problem. This may require considerable time and effort. At any point in the solution of an isolation problem more than one method or combination of methods can usually be employed. The choice of the most rapid and convenient procedure requires careful evaluation of many factors which include the available equipment, the time and assay facilities required for a particular step, use of previous knowledge or experience, and the scale of operation.

A major objective of this book is to provide perspectives for the commonly used methods and indications for their use. The determination of molecular properties useful in separation based on micro test methods, paper chromatography, thin-layer chromatography, and electrophoresis is described. Illustrative examples of each method are included.

Separations are classed generally as group or fractionation methods depending on the selectivity or fractionation needed. The theoretical principles of group-separation procedures, liquid–liquid partition, ion-exchange selectivity, gel permeation, and adsorption are included. Methods of influencing the selectivity coefficients are discussed.

The basic theory of fractionation methods is developed and the principles of application are discussed in terms useful to practicing chemists whose major interest is the use of separation methods to accomplish a given objective and not separation per se.

It is hoped that the book will serve as a useful guide to the solution of many practical problems encountered in the laboratory practice of organic chemistry and biochemistry.

I am indebted to many co-workers in the Merck Sharp & Dohme Research Laboratories who have offered many helpful suggestions during the preparation of the manuscript, especially L. Chaiet, R. G. Denkewalter, T. E. Jacob, E. A. Kaczka, A. J. Kempf, T. W. Miller, H. Shafer, and R. Weston. The

57187

cooperation of M. Tishler, L. H. Sarett, and H. B. Woodruff, through whose efforts many facilities of the Merck Sharp & Dohme Research Laboratories were made available, is gratefully acknowledged. I also wish to recognize the invaluable assistance of Mrs. M. M. Tatro for secretarial help in preparing the manuscript.

Rahway, New Jersey
May, 1969

FRANK J. WOLF

Contents

Introduction

I. TYPES OF SEPARATION CONSIDERED

Separation processes are based on mechanical and chemical methods. No separation can be achieved without some mechanical means. For the purpose of this book only separations in which the substances being separated are present in the orginal mixture in forms which cannot be physically separated are considered. Methods based on the degradation or decomposition of unwanted substances are likewise not included.

Consequently the separation methods of most interest are those which can be applied to solutions of two or more substances. During the separation process at least two phases are present and the separation is based on the unequal distribution of the substances between these phases. In solvent extraction and partition chromatography methods the phases are liquid. In other types of chromatography one of the phases is a solid.

II. DISTRIBUTION COEFFICIENT

A. Group Separation

The relative affinity of a substance for two phases is the distribution coefficient. This is frequently expressed as the ratio of concentration in the two phases. The ratio of the distribution coefficients for two substances is the separation factor, β, for the substances in the system under consideration. This is customarily expressed as a value greater than 1 rather than a fraction. If the β value is very large the separation is a "group" or type separation. If β is small a fractionation method is required. Thus, a group separation of acetic and propionic acid from glucose can be carried out by extraction of an aqueous solution with an immiscible solvent such as ethyl acetate. The further separation of acetic and propionic acids requires a fractionation method.

In the above example the group separation is that of a "polar" from a "nonpolar" substance. Separations may also be used to group materials based on their ionic properties. To a lesser extent molecular size can be used in grouping materials. Most separations of complex mixtures are carried out by using a group method at an early stage. Since the group separation step results in enrichment of related substances, one or more fractionation steps for the final separation of pure substances are usually required.

B. Fractionation Separations—Reasons for Using Them

Any separation method which produces a product of high purity in good yield requires either countercurrent contact of two phases or very high separation factors.

Suppose that two substances are present in equal amounts in a solution. A batch separation step is applied by introducing a second phase, either liquid or solid, which removes 91% of A and 50% of B. Since the distribution coefficients of A and B are 10 and 1, respectively, the separation factor, β, is 10. After the phases are separated one phase will contain 50% of B in 85% purity and the second phase will contain 91% of A in 64% purity. If the distribution coefficients do not change with the concentrations of A and B, it can be seen that there is no method of utilizing successive contacting of phases by which both good yields and purity can be obtained. This is true even though the starting material is 50% pure and a system having a β of 10 is used. Thus, for batch operation, even in successive stages, a much larger separation factor is needed if both high yield and purity are to be obtained.

If the second phase used in the above example is an immiscible liquid, the use of another solvent is not likely to markedly change the separation factor unless one of the compounds is affected by one of the solvents. Thus, methyl stearate and methyl oleate were observed to have a β of 1.6 ($K = 17.6$ and 10.8, respectively) in a system of 95% methanol–water–isooctane and a β of 2.6 ($K = 19$ and 7.3, respectively) in hexane–acetonitrile. The change in separation factor with the two systems is not of any great usefulness for group separation but may be significant for fractionation processes. Since the two solvent systems are rather different, the change in separation factor is about as large as can be obtained in the absence of some additional factor. It has observed that certain heavy metal salts, especially mercuric and silver salts, form association complexes with olefinic bonds. If mercuric acetate is added to the first solvent pair, the resulting complex with methyl oleate has increased polarity and the distribution coefficient is reduced from 10.8 to 0.28. Since methyl stearate is unaffected, a β of 63, suitable for group separation, is obtained. Other similar effects may be due to strong hydrogen bonding of one component in one system and not in another. A change in pH which influences the ionic state of one of the compounds also markedly affects the separation factor. If the distribution coefficient is not independent of concentration the separation factor of two compounds differs depending on the concentration of each and/or the amount of second phase employed. No predictions can be made for such systems in the absence of considerable experimental data.

C. Methods of Fractionation

If the separation requires substantial enrichment it is usually necessary that at least one step employ countercurrent contacting of two phases. Most laboratory countercurrent contacting methods are carried out by maintaining one phase stationary and moving the other phase past the stationary phase. In chromatography the stationary phase is solid. In countercurrent extraction techniques both phases of the system are liquids.

III. USE OF MICRO METHODS

The evolution of the rapid and simple techniques of paper and thin-layer chromatography and zone electrophoresis has been of great value in seeking methods or systems having adequate separation factors. Although these methods are frequently used as identification tools or in studies of purity, they can be also used to determine the chemical and physical properties of an unknown substance. The determination of differential migration rates of various substances under a variety of conditions can be carried out rapidly. The data obtained may be directly applicable in the design of a separation process, such as solvent extraction, ion exchange, or chromatography, and its large-scale application. In addition, these methods provide procedures for determining the behavior of the molecule in a complex environment. This study is usually necessary, even for known compounds which might be part of a synthesis reaction mixture, since the actual prediction of a "best" system based only on the structural features of the compound being isolated is uncommon.

These tests are useful in determining methods of carrying out both group and fractionation separations and in predicting the behavior of the desired compound in the separation step.

IV. EVALUATION OF SEPARATION PROCESSES

A separation method or process can be evaluated using four criteria. These are (1) yield, (2) separation, (3) capacity, and (4) efficiency. The advantages of having a high yield are obvious but if this is achieved with little separation the method is not satisfactory. Likewise if the separated product is obtained in low yield the method may be of little value. Some separation methods are readily applied on large scale and with large amounts of material; others can be applied on small scale only and hence would be less satisfactory on a capacity basis. Various criteria in the determination of efficiency are possible. Thus, excessive time, equipment, reagent, or labor cost may render a separation method impractical. An illustration of this is the analysis of amino acids by ion exchange chromatography. Although pure amino acids in a buffer solution are obtained from the chromatography column the method is not nearly efficient enough to be considered useful in the preparation of amino acids. The separation, although complete, is inefficient when equipment, time, and reagents are considered.

In summary, the overall separation of a pure substance from a complex mixture is a stepwise process employing a group separation and a fractionating separation. The group separation may be a batch process. The fractionation separation is a countercurrent process.

1 *General Principles*

I. STABILITY

Before attempting to determine which separation method is most applicable to a particular substance it is necessary to know conditions under which the compound is stable. A preliminary investigation should be undertaken to assess the effect of pH, solvent, temperature, light, and oxygen on the substance. Although some commercial separation methods can be carried out using a condition of great instability for the desired substance, these are based on detailed knowledge of the substance. Benzylpenicillin is rapidly decomposed at acidic pH. The rate of decomposition is influenced by both temperature and pH. However, a commercial process relies on the successful extraction of the free acid into solvent and reextraction into water as an alkali metal salt. High yields are obtained over these steps by using continuously flowing streams and centrifugal separators in such a way that the entire extraction and reextraction is complete in about 1 min. Such rapid transfers are usually not possible in ordinary laboratory equipment; hence it is desirable to avoid unstable conditions if possible.

Stability information is especially desirable in dealing with substances of biological origin. Frequently these materials are found to be unstable to drastic changes in pH or to some other laboratory stress. Fractionation steps may require prolonged contact and if the material is unstable may even result in lower, rather than higher, purity.

Some type of assay method is usually known for the desired substance. Useful stability information can be obtained using methods which do not require precise assay information if extended time or drastic conditions are used. Thus it may be more meaningful to determine the effect of pH on stability by using brief exposures to drastic pH changes than to attempt to determine rates of decomposition at close pH intervals. Similarly, temperature increases can be substituted for time provided conditions are chosen judiciously.

Although such stability studies require material, time, and assays and may appear to detract from the objective of obtaining pure material, the latter may be substantially aided by such information.

II. GROUP SEPARATIONS

Substances can be grouped based on ionic properties, polar properties, or molecular size. Each of these will be discussed in greater detail under the method employed. A general outline of the groups possible and the methods of obtaining these groups is given in this section.

A. Molecular Size

1. Dialysis and Ultrafiltration

The separation mechanism can be simply visualized as based on molecular size and shape. A membrane is used which has pores of sufficient size to allow smaller molecules to pass but which inhibits the passage of larger molecules. Thus in a batch process, two fractions are obtained. The exclusion properties of membranes vary drastically. Cellophane membranes can be specially treated to permit the passage of molecules as large as 20,000 m. wt. but the cellophane types commonly available are useful for separating substances of about 5000 m. wt. or lower from higher molecular weights. Special membranes have recently become available which are applicable to the separation of much lower molecular weight substances, using filtration. Reverse osmosis (high pressure filtration) can be used to separate salt from water using cellulose acetate films. Certain types of ion exchange membranes permit the passage of substances having molecular weights of less than 1000 but exclude higher molecular weight compounds.

2. Gel Filtration

The procedure consists of applying a solution containing substances of different molecular weights to a column containing a swollen gel in the desired solvent and developing the column with the solvent. The column can be considered as containing two types of solvent, that within the gel particle and that outside the gel particle. Large molecules, which cannot permeate the gel, appear in the column effluent after a volume equivalent to the solvent outside the gel has emerged from the column. Small molecules, which permeate the gel matrix, appear in the effluent after a volume equivalent to the total liquid volume within the column has emerged. A typical separation is depicted in Fig. I.1.

As a group separation method, the procedure can be used to separate substances which do not permeate the stationary phase from those which do.

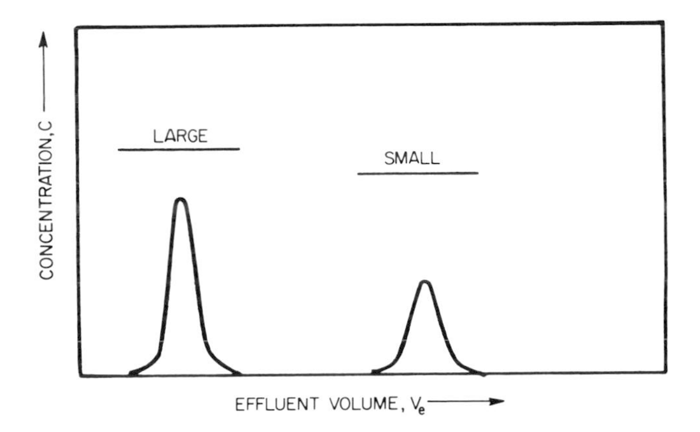

FIG. I.1. Group separation by gel filtration.

Separation of substances which partially permeate the gel may be incomplete depending on operating conditions.

The type of gel used in the column determines the size of the excluded substances. The variety of gels commercially available has continuously expanded. Hydrophilic gels having exclusion limits of as little as 1000 up to about 100,000 are now available and hydrophobic gels, which can be used in many organic solvents, are rapidly being developed. Glass particles having controlled pore size are also available. Thus by using gels it is possible to "size" molecules in solution into general groups depending on the types of gel employed. For example, a group containing all substances larger than about 1000 can be obtained by using a dense gel. This group could be resized on another gel to eliminate substances having a molecular weight greater than 10,000, yielding a fraction containing substances of m. wt. > 1000 but $< 10,000$, etc. The procedure may be difficult to operate if significant adsorption of high molecular weight substances is obtained.

The successful use of this method depends on careful attention to details of column packing and operation. These will be discussed in detail in Chapter IV.

B. Ionic Properties

1. Ion Exchange

This method can generally be applied to lower molecular weight compounds (up to 3000–5000) and separations based on dissociation constants can be used to group substances into weak or strong bases, weak or strong acids, amphoteric, and neutral fractions. The method is based on adsorption and/or elution from the ion exchange substances as a function of pH. For example, arginine, a strong base, adsorbs on cation exchange resins at all pH values up to 10, whereas, histidine, a weak base, is eluted at pH 8 but adsorbed at pH 5. Weak acids are adsorbed only at higher pH values and strong acids are adsorbed at lower pH values as well. Amphoteric compounds are not adsorbed at their isoelectric point but are adsorbed at higher or lower pH on the appropriate resin type. Neutral compounds are not adsorbed by ion exchange resins at any pH. As a general rule ion exchange is more applicable to polar compounds which are readily soluble in water than to nonpolar compounds of limited water solubility. Details of operation will be discussed later.

2. Solvent Extraction

Free and salt forms of acids and bases have markedly different distribution coefficients in liquid–liquid extraction systems. The salt form prefers the polar phase. This property can be used as a well known group separation method of acids, bases, and neutral compounds.

C. Polarity

1. Definition and General Principles

Group separations based on polarity are obtained by solvent extraction. The term "polarity" is generally not well defined and may have a variety of meanings. However, the concept is useful in considering the properties of organic molecules and permits the application of certain general principles of adsorption and solubility.

An accurate definition of polarity is difficult. Some correlation exists between polarity and the dielectric constant of a material in that highly polar substances can be expected to have high dielectric constants and highly nonpolar substances have low dielectric constants. The properties of a given molecule are a summation of its polar and nonpolar portions. Groups increasing polarity are ionized groups and groups capable of entering into hydrogen bonding either as a donor or acceptor or both. Groups incapable of entering into hydrogen bonding are nonpolar.

A partial listing of the dielectric constants of some common solvents is contained in Appendix I.

The well known rule of "like dissolves like" is especially applicable in considering the types of compounds which may dissolve in organic solvents. Apparent exceptions to this rule may occur as a result of hydrogen bonding capabilities or other molecular interactions. Thus $CHCl_3$ has been used extensively in the extraction and purification of alkaloids, possibly because of hydrogen bonding with weakly basic nitrogen atoms.

2. *Solvent Extraction*

If two phases are produced by a mixture of two or more solvents, one phase is more polar than the other. Considering systems containing only two pure solvents, various degrees of compatibility or mutual solubility of the solvents are obtained depending on the solvents employed. The less the degree of solubility the greater the polarity difference of the two phases produced. Thus extreme differences are obtained with hexane and water. On the other hand, *sec*-butanol or phenol and water produce two phases with much less difference in polarity. Compounds dissolved in a two-phase system distribute between the solvents on the basis of polarity similar to the solvent composition of the two phases. Thus, solvent extraction can be employed as a method of separation of polar and nonpolar compounds.

D. Summation of Group Properties and Methods

The general area of application of each of the grouping methods is illustrated in Fig. I.2. The boundaries between each group are somewhat indistinct and depend both on the type and method of carrying out the group separation as well as the properties of the desired compound. For the average mixture of compounds encountered in the separation of natural products it cannot be expected that any group method or combination of group methods

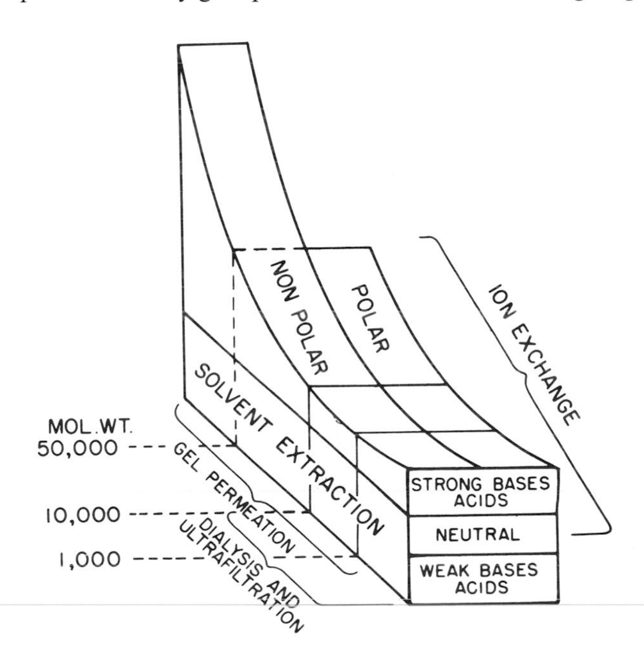

FIG. I.2. Summation of grouping methods.

will produce pure material. However, the grouping procedures eliminate many compounds and simplify the application of a final fractionation step.

III. FRACTIONATION SEPARATIONS

A. Requirements and General Methods

Fractionation is differentiated from group separation in a philosophical sense. If the separation factor is low, i.e., less than about 10, a fractionation method is needed. A satisfactory fractionation procedure should have the capability of separating materials with a separation factor as low as 1.3. This is usually sufficient for separating closely related compounds such as members of homologous series. However, structural isomers may have separation factors as low as 1.1. As previously mentioned, fractionation separations require a countercurrent procedure in which the differential rate of migration of the substances involved is used to effect a physical separation.

All countercurrent processes result in zone spreading and dilution. The product is always obtained from such a process in a greater solution volume than that used initially. Since liquid countercurrent distribution represents the simplest physical model and can be divided into discrete steps, it will be used to illustrate the basic principles of countercurrent processes.

B. Countercurrent Distribution

1. Procedure

The fundamental procedure is illustrated by the following example:

A solution of a substance is placed in the first of four vessels; an equal volume of immiscible solvent is added and the material is distributed by mixing the two phases. After equilibrium has been established the phases are separated. The upper phase is added to fresh lower phase in vessel number 2 and fresh upper phase is added to vessel number 1. After equilibration of the phases in the two vessels the upper phase from vessel number 2 is transferred to vessel number 3 which contains fresh lower phase, that from number 1 is placed in number 2 and fresh upper phase is added to number 1. After another series of such transfers, each of the four vessels would contain upper and lower phases.

The percentage distribution of material, x, having a partition ratio of 1 at each stage is illustrated by the boxes in Table I.1 and in Fig. I.3.

For a substance, y, having a partition ratio of 9, the distribution at each stage is illustrated by the numbers in parentheses in Table I.1.

Before considering a mathematical description of a countercurrent

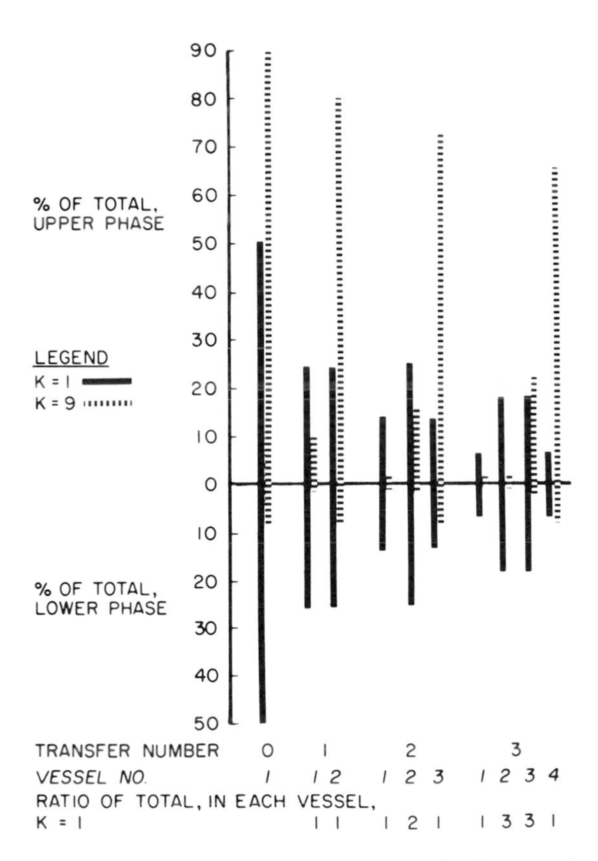

FIG. I.3. Stages of countercurrent extraction for four vessels.

process with many stages, the results of the above hypothetical case should be examined.

If substances x and y had been present in equal weights in the original mixture then the purity and yield of each substance in each phase of the four vessels would be as tabulated in Table I.2. Phases 2–lower and 1–upper and 1–lower contain 31.25% of substance x at a purity of 98.8%. By combining with fraction 2–upper and 3–lower, 68.75% of x can be obtained with a purity of 93% (fraction 2). Only one phase contains y in purity of 90% or greater and the yield is only 65.6%. By combining both phases of vessel 4, fraction 1 would contain 72.9% of substance y (purity 85.4%).

By continuing the process nearly quantitative yields of both x and y can be obtained in high purity. Apparatus useful for carrying out many transfers simultaneously will be described in Chapter III. Since as the number of transfers increases the absolute concentration decreases, it has been assumed

TABLE I.1

PERCENTAGE DISTRIBUTION OF MATERIAL x WITH A PARTITION RATIO OF 1 (BOXES) AND OF MATERIAL y WITH A PARTITION RATIO OF 9 (PARENTHESES)

Step No.	Vessel 1		Vessel 2		Vessel 3		Vessel 4		No. of transfers
1	(90%)	50%							0
	(10%)	50%							
2	(9%)	25%	(81%)	25%					1
	(1%)	25%	(9%)	25%					
Ratio		1		1					
3	(0.9%)	12.5%	(16.2%)	25%	(72.9%)	12.5%			2
	(0.1%)	12.5%	(1.8%)	25%	(8.1%)	12.5%			
Ratio		1		2		1			
4	(0.09%)	6.25%	(2.43%)	18.75%	(21.87%)	18.75%	(65.61%)	6.25%	3
	(0.01%)	6.25%	(0.27%)	18.75%	(2.43%)	18.75%	(7.29%)	6.25%	
Ratio		1		3		3		1	

TABLE I.2

SEPARATION OF TWO SUBSTANCES BY COUNTERCURRENT DISTRIBUTION (CCD)

	Vessel–Phase	Substance x		Substance y	
		Yield (%)	Purity (%)	Yield (%)	Purity (%)
Fraction 1	4–Upper	6.25	8.5	65.6	91.5
	4–Lower	6.25	43.0	7.3	57.0
	3–Upper	18.75	47.3	21.9	53.7
Fraction 2	3–Lower	18.75	88.6	2.4	11.4
	2–Upper	18.75	88.6	2.4	11.4
	2–Lower	18.75	98.8	0.3	1.2
	1–Upper	6.25	99	0.09	1.0
	1–Lower	6.25	99+	0.01	

that the partition ratio for each substance does not change appreciably with concentration. However, this assumption is not always valid. Factors influencing the partition ratio will be discussed in greater detail in later chapters.

2. *Parameters of the Separation Process—Relationship of Chromatography and Countercurrent Distribution*

For multistep countercurrent processes it is desirable to be able to calculate the following:

(1) The distribution ratio
(2) The number of vessels containing a certain yield of material
(3) The number of transfers needed to separate substances with known separation factors
(4) The number of effective stages for chromatographic techniques

Although chromatography differs from countercurrent distribution in several ways, the most important being the lack of equilibrium in the flowing system and the factors influencing mass transfer to and from the stationary phase, it can be considered as representing a stepwise process in which the solute passes from stage to stage and eventually emerges from the column or (paper and thin-layer chromatography) is located at a particular zone.

For a countercurrent extractor of cascade design in which a mobile phase contacts a stationary liquid phase and then moves on (countercurrent distribution, CCD) and a packed column containing a stationary phase the same relationship between elution peak, band spread, distribution ratio, and separation factors can be postulated.

By considering that the summation of all factors contributing to the chromatographic process is represented by the realized or effective zone occupied by a hypothetical plate the same mathematical treatment has been used to describe both processes. The classical papers of Martin and Synge (*1*) and Mayer and Tompkins (*2*) applied this treatment to partition chromatography and ion exchange chromatography. The fundamental difference between chromatography, which is a continuous process, and CCD, which is a discontinuous process, was pointed out by Glueckauf (*3*) and has been more fully developed in the recent works of Klinkenberg and Sjenitzer (*4*) and Giddings (*5*).

3. *Mathematical Description—Formation and Location of Bands*

A. Countercurrent Distribution. For CCD the number of theoretical plates or stages and the number of transfers is always known. For chromatography these can be determined from the observed band spread and retention volume.

Consider a multicelled extraction apparatus as illustrated in Fig. I.4 (*6*).

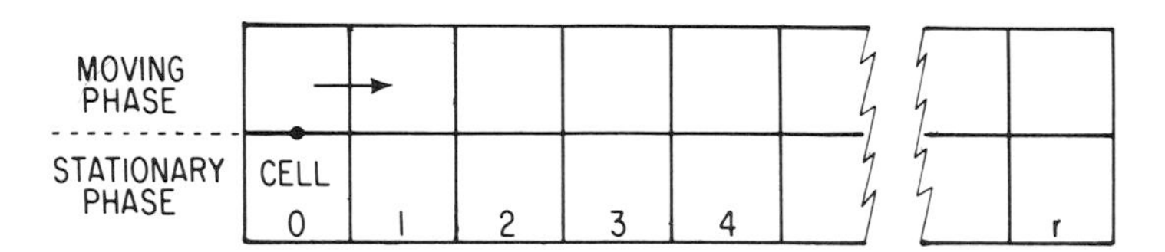

FIG. I.4. General countercurrent process.

A molecule is introduced into cell 0. The cell contents are equilibrated and a transfer step consisting of shifting the mobile phase one cell to the right and adding a new mobile phase into cell 0 is taken. If p is the probability that the molecule will be transferred in the mobile phase, $1 - p$ is the probability, q, that it will remain in the stationary phase. After two equilibration and transfer steps the probability that the molecule would have moved with the mobile phase both times is $p \times p$ and that it would have remained in the stationary phase both times is $q \times q$. For n transfers these probabilities are p^n, and q^n, respectively. At any time $(q + p)^n = 1$ and for n transfers the individual terms are described by the binomial expansion

$$q^n + np^{n-1}p + n(n - 1)q^{n-2}p^2 + \ldots, + p^n = 1 \tag{I.1}$$

In this case q^n represents the likelihood that the molecule will still be in the original cell after n transfers and p^n the likelihood of its having moved with each transfer. The probability, $T_{n,r}$ that the molecule will occur in a particular cell, r, after n transfers is given by the general expression for the rth term of the binomial expansion

$$T_{n,r} = \frac{n!\,q^{(n-r)}p^r}{r!\,(n - r)!} \tag{I.2}$$

After the mobile phase has reached the last cell of the apparatus, the cell (M) most likely to contain the molecule is

$$M = rp \tag{I.3}$$

Thus, if $p = 0.5$, M is $0.5r$, i.e., half-way through the apparatus, and n is $r - 1$.

Starting with many molecules of a pure substance instead of one, the distribution ratio, G, represents the ratio of the number of molecules in the moving phase to the number in the stationary phase in any cell. G is related to the partition ratio, K, by the relative volumes of mobile, V_m, and stationary phases, V_s, in any cell.

$$G = K(V_m/V_s). \tag{I.4}$$

Assuming that the value, p, for each molecule is not influenced by the presence of other molecules, then p represents the proportion of molecules in the moving phase

$$p = G/(G + 1) \tag{I.5}$$

The position of the cell M containing the maximum concentration as a fraction of the total number of cells, r, employed in the fractionation is M/r and is equal to R_f or relative mobility.

$$\frac{M}{r} = \frac{G}{G+1} = R_f \tag{I.6}$$

For countercurrent extraction both the number of cells, r, in the apparatus and the number of transfers, n, are known and the separation process can be examined rigorously.

B. CHROMATOGRAPHY. Chromatographic processes can be considered as having an "effective" number of transfers even though equilibrium is not established in a strict sense at each transfer. Thus Eqs. (I.4) and (I.6) can be applied in the determination of the partition ratio, K, and (I.3) determines the location of the zone center. These general relationships apply to partition chromatography, ion exchange chromatography, paper chromatography, gas chromatography, thin-layer chromatography, and gel permeation chromatography.

For column techniques in which the material is eluted from the column the peak concentration emerges from the column at an infinitesimal distance beyond the last cell, r, in the column and for all practical purposes

$$r = np \tag{I.7}$$

in which r is the total number of plates in the column and n is the total number of transfers required to transfer 50% of the material through the column. Let V_{pm} equal the volume of the mobile phase in one plate of the column; then the total mobile phase in the column, V_m, is rV_{pm} and the elution volume, V_e, is nV_{pm}. V_e is the volume which has emerged from the column up to the point of maximum concentration of the eluted substance. Multiplying Eq. (I.7) by V_{pm}

$$V_{pm}\, r = V_{pm}\, np \tag{I.8}$$

hence

$$V_m = V_e\, p \tag{I.9}$$

and

$$V_m = \frac{V_e G}{G+1} \tag{I.10}$$

In the case of partition chromatography substituting for G,

$$V_e = V_m + V_s/K \tag{I.11}$$

The elution volume, V_e, is seen to be independent of the number of plates in the column. It is determined only by the partition ratio, K, and the relative volume of mobile phase and stationary phase within the column. The relationships given in Eq. (I.9) were shown by Mayer and Tompkins (*2*) to apply to ion exchange chromatography of radioisotopes.

4. *Mathematical Description—Band Spread and Separation*

Thus far the location of the zone of maximum concentration has been determined in terms of the distribution coefficient and the volume of the mobile phase. This volume is independent of the number of transfers. However, no insight as to the number of transfers needed to effect a separation of two substances is provided by these relationships.

Since in countercurrent distribution there is a distinct separation of each transfer, this will be considered first. For operations up to 50 transfers the distribution of a substance in each tube can be obtained for different values of p from published probability tables. Such tables are especially useful and time-saving if the method is used for either the determination of p or purity determination by employing a small number of transfers. For operations having a total number of transfers greater than about 25 the concentration profile of a pure substance approximates a Gaussian distribution.

For the binomial expansion $(\mathrm{p} + q)^n = 1$, the standard deviation, or variance, σ, of the Gaussian curve is given by

$$\sigma = (npq)^{\frac{1}{2}} = \frac{(nG)^{\frac{1}{2}}}{G+1} \qquad \text{(I.12)}$$

A. PROPERTIES OF GAUSSIAN DISTRIBUTION. Since G is obtained by location of the maximum [Eqs. (I.3) and (I.5)], the calculated or theoretical curve can be quickly determined using published tables for the area under the Gaussian curve and fraction of maximum concentration for various values of σ. A summary of area and concentration profiles at various values of σ is included in Table I.3. The relationship of various features of the Gaussian curve is illustrated in Fig. I.5.

B. DEGREE OF SEPARATION OF PEAKS. *i. CCD.* The degree of separation of two substances can be calculated from the position of the maximum and the standard deviation, σ. Many procedures have been suggested for doing this. A general relationship which can be applied to either chromatography or CCD is as follows:

$$R_r = \frac{V_b - V_a}{(v\sigma_a + v\sigma_b)} \qquad \text{(I.13)}$$

R_r is relative resolution and V_b and V_a are relative volumes at the elution peak of substances A and B for chromatography or tube numbers for CCD. The volumes of one standard deviation, $v\sigma_a$ and $v\sigma_b$ for chromatography can be replaced by the number of transfers in one standard deviation for CCD.

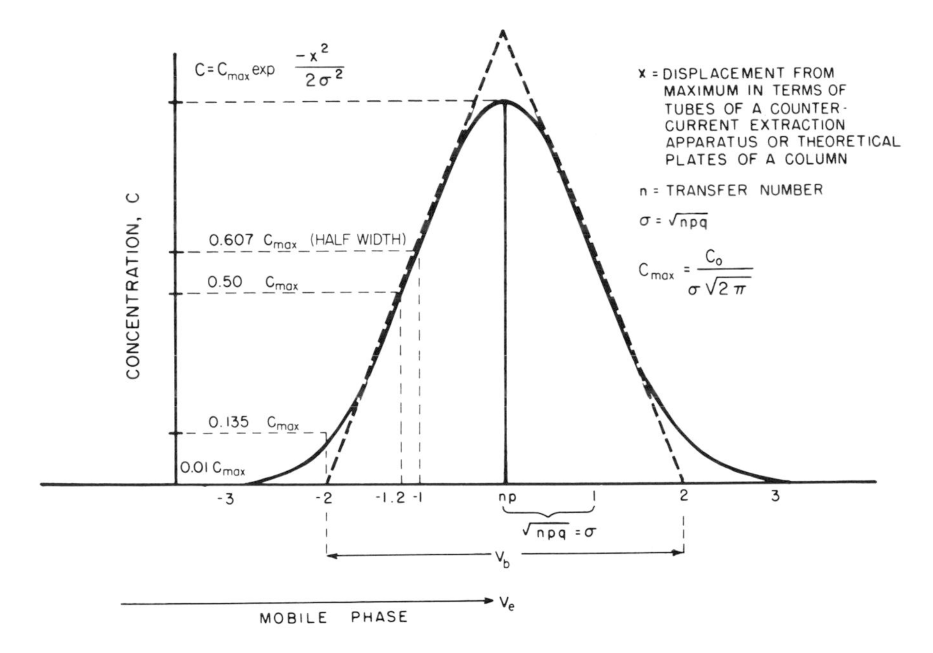

FIG. I.5. Features of the Gaussian curve.

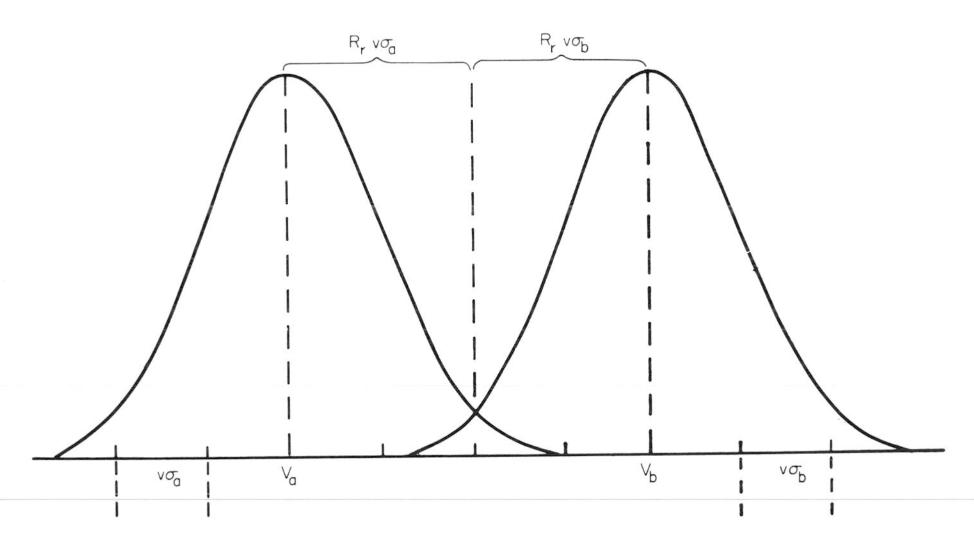

FIG. I.6. Resolution of two components, a and b.

TABLE I.3

PROPERTIES OF THE GAUSSIAN CURVE

σ	Area[a] (% of total)	Relative conc. (% of max.)	Absolute concentration for condition $n = 100$, $p = 0.5$ (fraction of feed concentration)
0	—	100	0.0797
0.5	19.15	88.27	0.07024
1.0	34.13	60.67	0.0482
1.5	43.32	30.32	0.02584
2.0	47.73	13.54	0.01077
2.5	49.38	4.39	0.00349
3.0	49.87	0.11	0.000878
3.5	49.98	0.022	0.00017
4.0	49.9999	0.002	—

[a] On one side of the point of maximum concentration, i.e., one-half the total area.

This is visualized in Fig. I.6. R_r is then the number of standard deviations contained in half of the two bands up to the midpoint of the overlap area. Thus if R_r is 2 there is 2.3% of each component in the overlap zone. Fraction I, collected up to 2σ beyond peak V_a, would contain a 97.7% yield of component A and fraction II begins 2σ before peak V_b and would contain 97.7% of component B. If R_r is 3 fraction I is collected up to 3σ beyond peak V_a and 99.9% of V_a would be contained in the fraction.

Since the location of the peak zone or tube is a function of the number of transfers for CCD or the volume throughput of a column containing a fixed number of plates and the standard deviation is a function of the square root of the same quantities, it is seen that the relative resolution increases only as the square root of the number of transfers or column height (plates).

The actual number of transfers required for a given degree of resolution can be calculated if the partition ratios are known by substituting values for the maximum [Eqs. (I.3), (I.5)] and the band spread [Eq. (I.12)] into Eq. (I.13).

$$n^{\frac{1}{2}} = R_r \frac{G_b^{\frac{1}{2}}(G_a + 1) + G_a^{\frac{1}{2}}(G_b + 1)}{G_b - G_a} \tag{I.14}$$

Returning to the first example of multiple extraction—the separation of two substances with partition ratios of 9 and 1 can be determined by substitution into Eq. (I.14), yielding

$$n^{\frac{1}{2}} = 2R_r$$

For 99.9% recovery of each compound, $R_r = 3$, therefore $n = 36$. It is to be noted that the minimum number of transfers is obtained when $G_a G_b = 1$.

If the partition ratios of two compounds are known, it may be advantageous to adjust the ratio of solvents or the mode of operation to conform to the above relation, all other factors being equal. Thus, for the above example if the distribution ratios are changed to 3 and $\frac{1}{3}$ merely by reducing the volume of mobile phase, then complete separation would be obtained in about 26 transfers. At this low number of transfers the Gaussian distribution on which this derivation is based is not exact.

ii. Chromatography—Determination of number of theoretical plates and separation. Although the concept of dividing a chromatographic column into discrete steps has been used profitably in the interpretation of chromatographic data, the model is not mathematically precise. One of the most glaring defects, that of assuming that the concentration of solute is constant in the mobile phase from the top to the bottom of a theoretical plate, was cited by Glueckauf in 1954 (*3*).

For continuously flowing systems the concentration profile of a substance emerging from a column is more precisely represented by a Poisson rather than a Gaussian distribution. This is the result of washout in the mobile phase and the fact that the material remaining in the column is subject to more transfers than the material which has emerged. The difference between the two concentration profiles, although predicted mathematically (*7*), is of little significance to the chromatographer. For columns containing as few as 40 theoretical plates an extremely accurate analysis is needed to detect any difference. For theoretical plate numbers greater than 100 the difference is mathematically indistinct. Thus for all practical purposes the concentration profile of the eluted zone is Gaussian.

The value of σ for CCD in terms of the number of cells in the apparatus is $(npq)^{\frac{1}{2}}$. For chromatography q, the fraction remaining in a cell (plate) after one transfer, is nearly 1 and σ in terms of plate numbers is approximated by $np^{\frac{1}{2}}$. Since in most chromatographic runs, the number of theoretical plates is large and p is small, the expression is adequate for most chromatographic work. For gas–liquid chromatography, Klinkenberg and Sjenitzer (*4*) and Giddings (*5*) have pointed out the contributions of diffusion rates, particle size, and fluid velocity on the number of theoretical plates in a column or chromatographic run. Although these are much different in absolute values for chromatography in all liquid systems, the overall effects are similar and a constant partition ratio produces a Gaussian curve. It is, therefore, not necessary to assess the effects of each variable in order to predict or analyze column behavior. Dixon (*8*) pointed out that the performance of chromatographic columns is readily determined from elution parameters.

For a chromatographic run, the elution record can be measured in terms of any convenient unit, such as time, distance, volume, or number of fractions. If the same unit of measurement is applied to the point of appearance of maximum concentration starting with the beginning of the run and to the band spread, the resulting dimensionless relationship can be used to determine the number of theoretical plates in the column for the eluted substance.

From Fig. I.5 it is seen that the intercept of the tangents of the Gaussian curve is 4σ. The dimensionless expression for the point of maximum concentration [Eq. (I.7)] is

$$n = r/p$$

and the number of plates per σ is $(1/p)r^{\frac{1}{2}}$.

Letting N equal the number of transfers in 4σ

$$N = (4/p)r^{\frac{1}{2}} \tag{I.15}$$

and

$$4n/N = r^{\frac{1}{2}} \tag{I.16}$$

The dimensions of a theoretical plate can then be used as needed. For example, if V_{pm} is the volume of a theoretical plate, V_e, the elution volume is nV_{pm} and V_b, the 4σ band spread is NV_{pm}. Hence,

$$16(V_e/V_b)^2 = r \tag{I.17}$$

This is the well known expression used for the determination of the number of stages in a column. The same fundamental result, derived and expressed differently, was obtained by Glueckauf (*3*). This measurement can be applied to columns operated under conditions in which p is constant (or nearly so). Gradient elution procedures produce drastic changes in p, and band spread measurements cannot be used to evaluate column performance.

iii. Factors affecting plate height. Present chromatography theory, based largely on gas–liquid chromatography systems, has resulted in the experimental evaluation of factors other than the partition ratio which influence the zone spreading. For liquid chromatography, the diffusion coefficient in the immobile phase is frequently an important factor in determining the height of a theoretical plate. If the immobile phase is a solid, this is a function of particle size and uniformity, and maximum performance is obtained using the smallest particle sizes practical. For chromatography of amino acids this factor has been studied by Hamilton *et al.* (*9*). Flow rate limitations with extremely small particles may cause practical problems.

iv. Effect of the distribution coefficient on the operation of columns. Although, as for CCD, the maximum resolution in terms of the number of transfers needed to separate two substances with a given β, reaches a minimum at the point at which the product of the distribution coefficients of the two substances is 1, more transfers are obtained per unit column length (assuming constant plate height) as the distribution coefficient becomes lower. The net effect is an increase in resolution as the distribution coefficient decreases. Carpenter and Hess (*10*) pointed out that below $G = 0.1$ the increase is negligible. A similar conclusion was reached by Dixon (*8*).

The following examples illustrate the effects of changes in the distribution coefficient assuming no change in β, plate height, or stationary phase/mobile phase ratio. Two substances having a β value (G_b/G_a) of 1.25 are considered in four systems in which substance A has a distribution coefficient of 0.1, 0.2, 0.8, and 2. G_aG_b is therefore 0.0125, 0.05, 0.8, and 5.0, respectively.

The number of theoretical plates required for a certain resolution of two materials by chromatography obtained by substituting Eqs. (I.7) and (I.16) into (I.13) is as follows:

$$r^{\frac{1}{2}} = \frac{R_r(p_a + p_b)}{(p_a - p_b)} \tag{I.18}$$

Examples calculated for $R_r = 2$ are contained in Table I.4.

TABLE I.4

EFFECT OF DISTRIBUTION COEFFICIENT ON COLUMN OPERATION

G_a	G_b	r	n	Volume of rich cut: Fract. of V_m	Volume of rich cut: Relative[a]
0.1	0.125	400	4000	2.2	880
0.2	0.25	484	2420	1.09	530
0.8	1.0	1156	1445	0.26	302
2.0	2.5	3364	1682	0.103	346

[a] Assuming constant volume and height per theoretical plate.

If it is assumed that the flow rate and volume per column unit is the same, then the time and volume of solvent required are directly related to n, the total number of transfers. Column length and volume are proportional to r.

In considering the maximum performance, factors such as time, dilution of the product, and total solvent requirements lead to different maxima. The calculations also indicate the desirability of determining the distribution coefficient before attempting chromatography. Low distribution coefficients inevitably require excessive time and solvent, whereas high distribution coefficients require longer and more efficient columns.

The assumption of constant flow rate per theoretical plate may be difficult to achieve in cases of finely divided adsorbent and tall columns. Since decreasing the distribution coefficient decreases the height needed to effect the same separation, less saving in time may be obtained in actual practice than the above table indicates.

v. Effect of changing distribution coefficient. For many adsorption processes and in certain liquid–liquid processes the distribution coefficient changes with concentration.

Elution profiles obtained will be sharp on the leading edge (p increases with decreasing concentration) with a diffuse trailing boundary or sharp on the trailing edge (p decreases with decreasing concentration) and diffuse on the front boundary. Although many useful separations can be achieved under these circumstances, the analysis of the separation may be impossible.

Factors influencing the distribution coefficient which contribute to non-linearity and methods of varying the distribution coefficient will be discussed under the individual methods.

REFERENCES

1. Martin, A. J. P., and Synge, R. L. M., *Biochem. J.* **35**, 1358–1368 (1941).
2. Mayer, S. W., and Tompkins, E. R., *J. Am. Chem. Soc.* **69**, 2866–2874 (1947).
3. Glueckauf, E., *Trans. Faraday Soc.* **51**, 34–44 (1955); Ion Exchange and its Applications, *Soc. Chem. Ind.* (*London*) pp. 27–38 (1954).
4. Klinkenberg, A., and Sjenitzer, F., *Chem. Eng. Sci.* **6**, 258–270 (1956).
5. Giddings, J. C., "Dynamics of Chromatography. Part I: Principles and Theory." Dekker, New York, 1965.
6. Connors, K. A., and Erikson, S. P., *Am. J. Pharm. Ed.* **29**, 509–517 (1965).
7. Brenner, M., Niederwieser, A., Pataki, G., and Weber, R., *in* "Thin-Layer Chromatography" (E. Stahl, ed.), English ed., pp. 75–133. Academic Press, New York, 1965.
8. Dixon, H. B. F., *J. Chromatog.* **7**, 467–476 (1962).
9. Hamilton, P. B., Bogue, D. C., and Anderson, R. A., *Anal. Chem.* **32**, 1782–1792(1960).
10. Carpenter, F. H., and Hess, G. P., *J. Am. Chem. Soc.* **78**, 3351–3359 (1965).

II *Determination of Molecular Properties*

I. INTRODUCTION

The successful separation of an unknown substance is greatly facilitated if its chemical and physical properties are known. For compounds of known structure, many of these properties can be approximated based on relationship to known compounds and the types of functional groups present. For unknown substances studies with paper chromatography, thin-layer chromatography, paper or gel electrophoresis, and small-scale tests carried out with solvents and ion exchange resins may reveal many important features of an unknown material even though only small amounts of crude material are available. These methods are particularly useful for determining the ionic properties, polarity, and molecular size. Subsequently solvent extraction, ion exchange, or adsorption methods can be used on large scale for the preparation of concentrates of the unknown material.

This chapter is concerned primarily with the use of these test methods and is not intended to be an extensive review of the techniques and details. Many good reviews of these are available to the reader and applications to many types of compounds have been described. Frequently, the objective of micro methods is the identification or detection of a known material. The objective of the present discussion is to point out how the information obtained from these tests may be utilized in the design of an effective separation process for an unknown substance. Since at any stage in an isolation problem there is usually more than one alternative, selection of the most efficient step diminishes the time and effort required and may even make the difference between success and failure. The relationships of micro test methods with processing steps for either group or fractionation separation will be discussed.

II. CHROMATOGRAPHIC PROCEDURES

A. Visualization

To apply the microchromatographic methods some procedure for the visualization of the compound must be available. Ideally this should be specific for the desired compound and sensitive to small amounts of material (< 10 μg). However, many nonspecific tests can be used in a screening fashion if some secondary test is available for occasional checking. At later stages in an isolation problem nonspecific tests are also useful for detecting impurities in the product.

Chromatograms may be visualized by direct observation, by the use or chemical tests, by the use of biological tests, or by effects on enzymatic or chemical reactions.

1. Direct Visualization

Substances with intense visible or ultraviolet light absorption and fluorescent substances may be visualized directly on paper strips. Thin-layer plates may also be used but are not as readily observed. Some thin-layer adsorbents contain fluorescent materials or fluorescent substances may be sprayed on the chromatogram. Ultraviolet light which excites the fluorescence may be absorbed by substances in the plate or on the strip. These materials cause the appearance of dark zones.

Viewing of chromatograms for ultraviolet absorption is aided by using a black box fitted with a method of passing ultraviolet light through the chromatogram. Light-absorbing zones are revealed by placing a phosphor-coated plate on top of the chromatogram. Fluorescence is observed by using reflected, filtered ultraviolet light and viewing in a dark box. Nonfluorescent ultraviolet-absorbing zones may be observed in this manner also but with lowered sensitivity.

Fluorescence is especially sensitive to pH changes and exposure of the developed chromatogram to alkaline and acidic vapors may reveal zones otherwise undetectable.

Direct visualization of hydrophobic substances on silica thin-layer plates can sometimes be accomplished by spraying the plate with water. The difference in wettability of the silica reveals the presence of the compound.

2. Chemical Tests

The development of color, ultraviolet light absorption, or fluorescence by chemical treatment is a detailed subject and has been successfully employed with varying degrees of specificity for many compounds. A list of some of the most universally applicable (and hence less specific) reagents is included in Appendix III and they will not be discussed here. Thin-layer plates with chemically inert adsorbents allow a wider selection of tests than paper strips.

In carrying out chemical tests the reagents are usually applied to the chromatogram by a spray technique. The reagent concentration should be such that a uniform light spray can be applied without danger of distorting or dislocating the developed chromatogram. If multiple tests can be applied to the same chromatogram, this may aid in selecting the desired zone. For

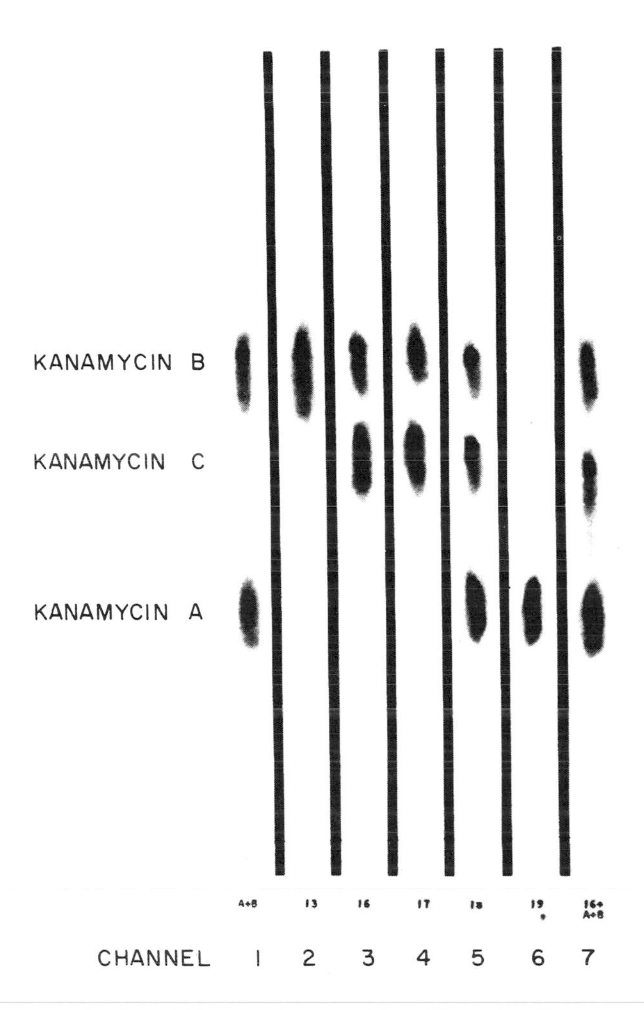

FIG. II.1. Visualization of Kanamycin by staining. Channel 1, Kanamycin A and B; Channel 2, column cut 13; Channel 3, column cut 16; Channel 4, column cut 17; Channel 5, column cut 18; Channel 6, column cut 19; Channel 7, mixture of 1 and 3.

example, zones located with iodine vapor can be marked and tested with more specific reagents. Ninhydrin sprays do not prevent the subsequent application of diazo coupling reagents, or even, in some cases successful bioautographs. This multiple test technique is valuable in pinpointing desired compounds present in a mixture under conditions for which superposition of the zones might be difficult.

Certain staining techniques in which acidic or basic dyes are applied to the chromatogram and then washed under conditions which remove soluble dye but not that fixed to compound have been quite successful in locating polyvalent compounds. Although this procedure has been used most extensively for proteins it can also be used for low molecular weight materials if the proper solvents are used. Thus kanamycin, an antibiotic with four basic groups, can readily be visualized with certain acidic dyes. This staining was used to follow the separation of three kanamycin types (Fig. II.1) (*1*).

3. Bioautography

If the substance being tested has activity in a biological system, this type of test usually has the needed specificity. For substances affecting the growth of microorganisms the paper strip (or a paper strip "print" of a thin-layer chromatogram) is placed on a properly prepared solid medium. For growth inhibition the medium is seeded with a sensitive organism. For growth stimulation an organism requiring the substance for growth is used. The medium is then incubated leaving the paper strip in place or removing it after allowing diffusion into the medium.

This technique is admirably adapted to the detection of antibiotics since the substance being sought has usually been observed to influence the growth of a microorganism in a solid medium. Usually only small amounts of active

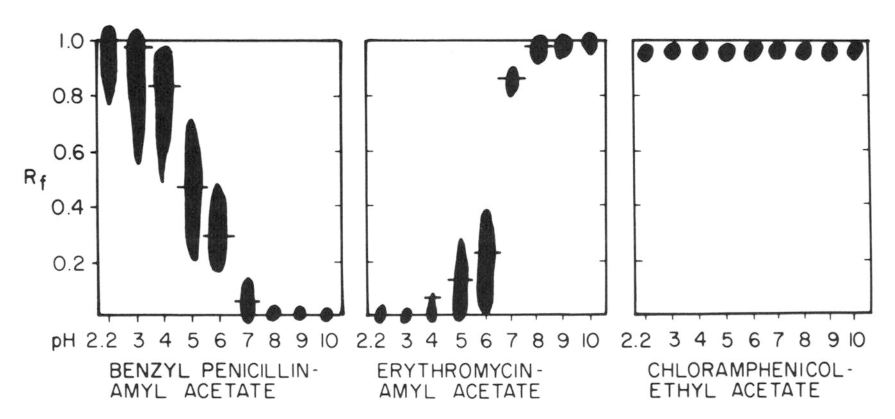

FIG. II.2. Bioautographs of antibiotics. Effect of pH on mobility of acidic, basic, and neutral antibiotics.

substance are needed. The effect of pH on the paper strip mobility of antibiotics has been shown to be a useful aid in early studies (*2*) (Fig. II.2).

If a direct test cannot be used because of limitations of the test conditions, the chromatogram can be sectioned and each section eluted and tested. For potent compounds the latter technique can be used for tests in animals or with plants.

4. Reactivity or Effect on a Test System

Reactivity or effect on the reactivity of a test system has been used in many ingenious ways. An example is the detection of acetylcholinesterase-inhibiting insecticides (*3*). To carry out this test the developed chromatogram is placed in contact with a strip of wet paper of the same size. A second strip, dipped in a solution containing acetylcholinesterase, buffer, and an acid–base indicator is then added. A final strip containing acetylcholine is then added to the stack. After a period of time the background of the chromatogram changes color due to the release of acetic acid. Areas containing the enzyme inhibitor remain unaffected. The use of paper impregnated with reactants allows more accurate control of reagent quantities and the application is more uniform than that obtained by spraying.

Tests employing diffusion through a solid medium such as agar, while commonly used for growth of microorganisms, can also be used with a variety of test systems. Tests of this type have the advantage of ready availability, speed, and reproducibility. The types of tests which can be adapted to solid media are limited only by the imagination of the investigator. Reagents can be added to the agar originally or diffused by flooding the agar surface with liquid or spraying. Known quantities of reactants can be added to paper and layered on the surface. The timing of addition of reagents can be effective in quantitating this response.

These tests can be used to detect the generation of fluorescence, areas of combination with visualization reagents such as protein stains, or color liberation, such as in the phosphate test.

Techniques of this type may have high sensitivity and, because of enzyme specificity, permit detection of the desired substance even though the resolving power of the chromatogram is limited. However, the physical properties of the detected material can be studied even though only very impure materials are available.

III. MICRO TESTS

A. Type and Sequence of Tests

1. Solvent Extraction

Micro tests useful in determining the physical properties of an unknown substance can be carried out using the minimum amount of material which can be assayed. Relatively few assays are required to yield much information. Initial tests do not require high precision in the assay. A second series of tests may be advisable using the information obtained in the first series as a guide to the next, etc. The following tests have been found useful in the preliminary screening and quickly yield information of a general nature.

Solvent extraction of an aqueous solution is first carried out at three widely separated pH values (pH 2.8–3.2, 7, and 9.5–9.8) using a polar and a nonpolar solvent. One or two extractions with *n*-butanol and chloroform or ethyl acetate and determination of nonextracted material are frequently sufficient. Highly polar materials may not be extracted at any pH. Less polar materials are extracted at alkaline pH if basic or at acidic pH if acidic. Substances extracted at all pH values with the nonpolar solvent should be tested subsequently with solvents of even lower polarity, such as *n*-hexane or benzene.

Indications that the material is acidic or basic should be followed up with extractions at closer pH spacing and with more precise assays. Very strongly polar substances are sometimes extracted with phenol but generally solvents of greater polarity than *n*-butanol are difficult to handle.

2. *Ion Exchange*

Polar compounds which are ionic in nature are removed by ion exchange resins. This test is conveniently carried out on a small scale by passing a dilute aqueous solution slowly through a bed of ion exchange resin. About one bed volume of liquid should pass through the column in 10–15 min. Several bed volumes should be used. The preliminary screening should be carried out with a strong acid resin ($-SO_3^-Na^+$) and a strong base resin [$-N(CH_3)_3^+Cl^-$] of low cross-linking. The cation exchange test should be run at pH 7 and pH 2. The anion exchange test should be carried out at pH 5 and pH 9.

Secondary tests can be carried out using intermediate pH values and weakly acidic or basic exchangers if indicated.

B. Compound Groups

As a result of these tests 12 groups of compounds can be clearly distinguished. These are polar or nonpolar, weak bases, strong bases, weak acids, strong acids, amphoteric, and neutral compounds. Compounds of intermediate polarity and of intermediate acidity or basicity may be more difficult to recognize by this procedure. These are best studied by chromatographic methods if possible.

C. Determination of Approximate Dissociation Constant and p*K*

1. *Introduction*

Knowledge of the p*K* of ionic compounds is generally useful in determining optimum conditions for either extraction or ion exchange processes.

Fortunately, this can be approximated by several independent procedures. These are generally based on the Henderson-Hasselbach equation showing the relationship between pH, pK, and the degree of ionization:

$$\mathrm{pH} = \mathrm{p}K + \log \frac{\text{ionized}}{\text{nonionized}} \tag{II.1}$$

As a general rule the ionized form of an organic compound does not extract into organic solvents as well as the nonionized form. Solvent extraction data expressed as the fraction extracted at different pH values can be used to approximate the pK. Similarly paper strip or thin-layer mobility at various pH values can be used to determine pK. Unfortunately both of these methods are most accurate using buffers which provide a constant effect on the distribution coefficient over a sufficiently wide range. The pK of weak or intermediate strength acids and bases can be estimated by ion exchange methods by determining the effect of pH on the distribution ratio. This method is approximate since it is necessary to assume no interaction between the resin backbone and the adsorbed material. This assumption is rarely valid for organic compounds. A fourth method of approximating pK is by determination of electrophoretic mobility at a variety of pH values. This method can be rather precise if precautions are taken to minimize effects due to adsorption to the support medium and electroosmosis. Fortunately, it is possible to correct for these effects.

Each procedure is discussed in some detail.

2. *Solvent Extraction*

For solvent extraction the partition coefficient in a solvent–aqueous system is determined (see Chapter III) at a series of pH values so that two or three points are obtained with 3 pH units on each side of the pK. One solvent pair is used. The estimation of the pK value from the data obtained depends on the following assumptions:

(1) The buffer effect does not vary with pH

(2) The solute does not form pH-dependent "association complexes" with any component of the system

(3) The ionized form is not extracted into the solvent phase

(4) The compound contains only one type of ionizable group

Although these assumptions are rarely completely valid, the pK for many compounds can be estimated within 0.5 units if the system is chosen to minimize deviations. Thus, buffer salts which depend on extractable organic acids or bases should be avoided and only mineral acids and bases used if at all possible. Buffer concentrations should be constant. Solvents with a minimum of hydrogen bonding potential should be used. The lowest practical concentration of substrate should be employed.

The relationship between the partition coefficient, pK, and pH is seen from the following:

Let K be the partition ratio C_s/C_w, for the nonionized form; K_d, the dissociation constant; α, the fraction ionized; and K_{eff}, the effective (observed) partition ratio. Then

$$K_{eff} = K(1 - \alpha)$$

and for acids

$$\alpha = \frac{(A^-)}{[(HA) + (A^-)]}$$

Therefore,

$$K_{eff} = K\left[1 - \frac{(A^-)}{(HA) + (A^-)}\right] = K\left[\frac{(HA)}{(HA) + (A^-)}\right]$$

Substituting for (HA) from $(HA) = (A^-)(H^+)/K_d$

$$K_{eff} = K\left(\frac{(H^+)}{(H^+) + K_d}\right) \tag{II.2}$$

which can be arranged to

$$\frac{1}{K_{eff}} = \frac{1}{K} + \frac{K_d}{K(H^+)} \tag{II.3}$$

or, expressed in logarithmic form,

$$\log K_{eff} = \log K - \text{pH} - \log[(H^+) + K_d] \tag{II.4}$$

At the pH one-half value, $(H^+) = K_d$ and

$$K_{eff} = K/2 \tag{II.5}$$

For bases (OH^-) is substituted for (H^+).

The calculated curve for K_{eff} vs. pH is plotted in Fig. II.3. Since the basic equation was derived by assuming that the ionized form has *no* solvent extractability the curve can only approach 0 as the pH increases. Actually, many ionized forms have a small partition ratio into the solvent and the resulting experimental curve may be S-shaped, as indicated by the dotted line in Fig. II.3. The actual determination of partition ratio requires assays of good precision. Typical data for crude gibberellins are illustrated in Table II.1.

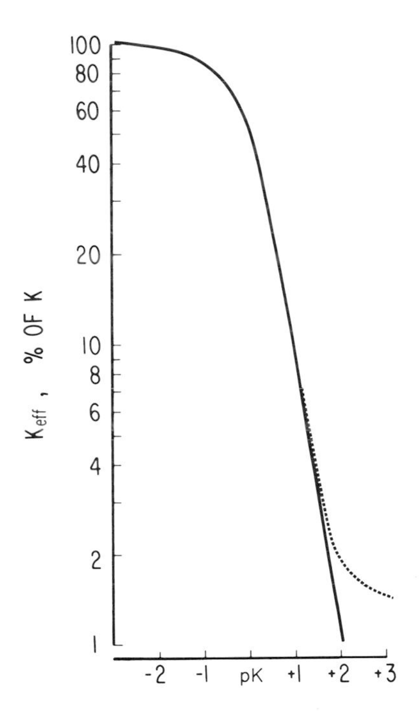

FIG. II.3. Effect of pH on partition ratio.

TABLE II.1

PARTITION RATIO OF GIBBERELLIN MIXTURE, SOLVENT/WATER

pH	Amyl acetate[a]	Ethyl acetate
3	4.85	3.54
4	3.16	3.0
5	1.5	1.0
6	1.0	0.18
7	0.81	0.05
Extrapolated pK	4.3	4.7

[a] Contained about 15% amyl alcohol.

Examination of these data indicates much greater distribution of the salt form into amyl acetate than into ethyl acetate. The method was applied to a variety of phenols by Golumbic *et al.* (*4*).

Similarly the data can be utilized in the reciprocal form, Eq. (II.3). This is illustrated for some pyridine alkaloids by Badgett *et al.* (*5*).

3. Paper and Thin-Layer Chromatography

The R_f is determined at a series of pH values using one solvent system, In addition to the assumptions listed for solvent extraction it is also assumed that the support for the system does not influence the R_f differently at different pH values. Under these conditions the R_f is a function of the partition ratio of the substance between the mobile and stationary phases. Systems having either the polar or nonpolar phase stationary may be used. For nonpolar compounds, "reversed" phase systems in which the support is impregnated with a nonpolar compound, such as Vaseline, decalin, or a higher alcohol may be advantageous. In general the polar phase should contain only one type of counterion, which should be highly polar and strongly ionized.

The relationship between R_f and K has been shown earlier [Eqs. (I.4) and (I.6)]. Assuming that the ratio of V_m/V_s does not change during the course of the development of the paper strip then the relation

$$G = \frac{R_f}{1 - R_f}$$

describes the R_f at a particular pH. From Eq. (II.2) the effect of pH on the R_f can be determined.

$$\frac{G_{\mathrm{eff}}}{G} = \frac{(\mathrm{H}^+)}{(\mathrm{H}^+) + K_d} = \frac{(R_{f,\,\mathrm{eff}})(1 - R_f)}{(1 - R_{f,\,\mathrm{eff}})R_f} \qquad \text{(II.6)}$$

In this relationship R_f is the maximum R_f obtained under any circumstance and $R_{f,\mathrm{eff}}$ is the observed R_f at a particular pH. This relationship for different values of G was used by Waksmundski and Soczewinski (*6*) to calculate a series of theoretical curves for different partition ratios. These are reproduced in Fig. II.4. The pK is conveniently determined by plotting the observed R_f values on tracing paper and superimposing the curve on the theoretical curve. The pK is then approximated directly. No fit will be obtained if the compound contains two ionizable groups of different pK or if the buffer or support system interfere. Using the data of Fig. II.2., the approximate pK_a for penicillin is found to be p$K_a = 3.0$ and for erythromycin p$K_b = 9.25$. The true values are reported to be 2.7 and 8.8, respectively. This method can be applied to pH chromatograms with widely varied R_f values. It may be more accurate if a system giving an R_f value of 0.4–0.9 for the nonionized form is used. Under these circumstances, Eq. (II.6) can be expressed logarithmically and gives the following relationship at the pH = pK value.

$$\log\left(\frac{1}{R_{f,\,\mathrm{eff}}} - 1\right) = \log\left(\frac{1}{R_f} - 1\right) + \log 2 \qquad \text{(II.6a)}$$

These methods, while admittedly not accurate or always applicable, are

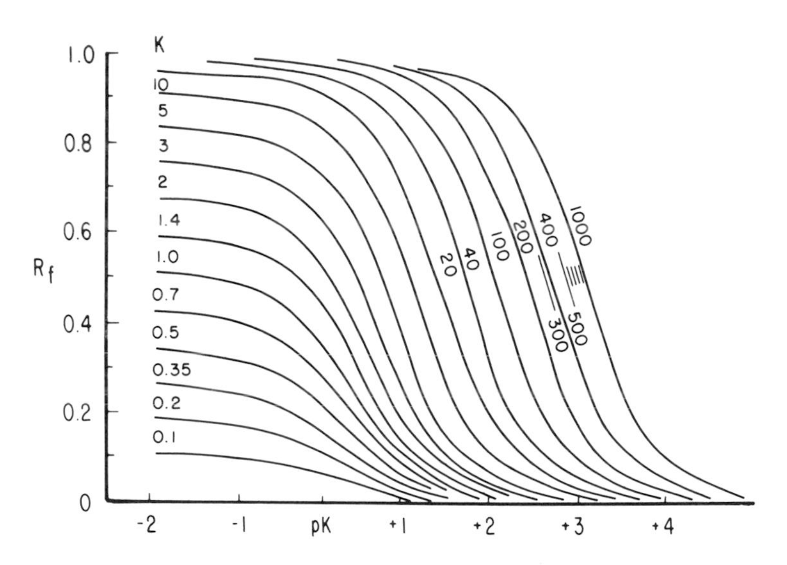

FIG. II.4. Effect of pH, pK, and partition ratio on R_f.

remarkably simple procedures which yield information concerning the substance being investigated with a minimum of effort.

4. *Ion Exchange*

For strongly polar substances which are not solvent-soluble, ion exchange reactions can sometimes be used in an indirect determination of ionization constants lower than about 10^{-3} using crude mixtures. Although the pH one-half value cannot be determined with an accuracy of greater than ± 1 pH unit, the procedure outlined below may yield valuable group separation information. No interpretation of the pH one-half value is possible with multivalent substances or with those compounds having ionization constants greater than about 10^{-3}.

Since there is no quick and ready method of differentiating the contributions of the nonionic interaction of resin backbone and the compound, best results are obtained if this is minimum. Aromatic compounds may have a high attraction for polymer backbone.

The following assumptions are necessary:

(1) The degree of ionization of the resin and the dominant counterion must be constant, preferably nearly 100%, at all pH values of the experiment.

(2) Nonionic forces influencing the distribution coefficient must be independent of pH and a relatively minor fraction of the total distribution coefficient.

(3) The distribution coefficient must be linear with concentration over the range employed.

(4) Complexes which affect the fraction of ionized substance are either not present or are not influenced by pH.

(5) The compound being studied is small enough to permeate the gel phase.

Consider the ion exchange reaction of an organic base with a sulfonic acid cation exchanger, $RSO_3^- \ M^+$.

$$[RSO_3^- \ M^+] + B^+ \rightleftharpoons [RSO_3^- \ B^+] + M^+$$

$$K_{sel} = \frac{[RSO_3^- \ B^+](M^+)}{[RSO_3^- \ M^+](B^+)} \qquad (II.7)$$

K_{sel} is an arbitrary "selectivity" coefficient. For convenience and simplified experimental conditions it can be expressed in terms of concentration per unit volume. Since the resin phase volume under the conditions specified remains nearly constant, it is not necessary to determine the absolute concentration of charge groups. This procedure, although not particularly satisfying to the purist, is adequate for the purposes of this discussion as long as the same sample of resin is used for the entire experiment. Bracketed quantities [ ] represent the resin phase and those in parentheses ( ), the solution phase. Substitution for K_d yields

$$K_{sel} = \frac{[RSO_3^- \ B^+]}{[RSO_3^- \ M^+]} \frac{(M^+)(OH^-)}{(BOH) \ K_d}$$

which can be rearranged and expressed in logarithmic form

$$\log K_d = \log \frac{[RSO_3^- \ B^+]}{(BOH)} + \log \frac{(M^+)}{[RSO_3^- \ M^+]} - \log K_{sel} - p(OH) \qquad (II.8)$$

Consideration of the above expression indicates the experimental protocol and reemphasizes the multiple corrections needed if the assumptions are not valid. The distribution coefficient $[RSO_3^- B^+]/(BOH)$ measured under conditions of high salt or buffer concentration and with a large excess of resin is expected to be relatively constant. In addition if greater than 90% of the resin remains in the original state the quantity, $\log (M^+)/[RSO_3^- M^+]$, is nearly constant. If the buffer concentration is about the same as the total concentration of ionized sites within the resin the term approaches 0. If no highly specific interaction with the backbone exists then $\log K_{sel}$ (expressed as the ratio of concentration of B in the solution and resin phases) is probably 0.0 to 1.5, and may be approximated by assuming 1. Thus, if the distribution coefficient is determined at a series of pH values at about 1.5 to 2 pH intervals, at constant ionic strength and at low loading, determination of the 50% absorption value by extrapolation from a logarithmic plot yields directly the log K_d after correction for the assumed K_{sel}. Since greater interaction with the backbone causes an increase in K_{sel} the K_d value obtained would be high if a large K_{sel} existed. Since phenylalanine has a K of less than 50 in 0.2 N buffer (pH 3.25) using polystyrene-based resin, the expectation that the correction due to this factor will usually be within 1 log unit for compounds which are not solvent-soluble is probably justified.

For acidic compounds the same derivation can be employed and a strong base, quaternary resin would be used.

The precise determination of the distribution ratio depends on the reliability of the assay method. The ratio can be calculated in any convenient unit since the final data are handled in a relative fashion. The determinations should all be performed under similar conditions of temperature and time. At the low resin loading required column procedures are not applicable and a batch method must be used. Sufficient time for the attainment of equilibrium (at least 30 min) should be allowed. To avoid possible exclusion effects of large molecules, only highly porous resin phases should be used with unknown compounds.

5. *Electrophoresis*

The migration of a substance under the influence of an electric field is a direct method of determining the net charge of the material in the environment of the experiment. It is readily apparent that an ion bearing a negative charge will be attracted to the positive electrode (anode) and an ion bearing positive charge will be attracted to the negative electrode (cathode).

This simple concept, which provides an absolute and direct relationship between the degree of charge and strength of the electric field, has been the subject of investigation for many years. Since the formation and existence of ions for most compounds usually requires the presence of water, measurements of net charge are usually carried out in water solutions. Many types of equipment and methods have been devised for determining electrophoretic behavior. It is beyond the scope of this book to review the many methods which have been employed. Successful application of the method requires that the apparatus employed be capable of transferring an electric field to the environment of the sample being studied and that some method of determining the net effect of the electric field on the desired compound be available. The experimental problems are more difficult than would seem at first glance. The major problem is due to the difficulties of preventing convection currents in the medium from destroying the separation. A second problem arises from the fact that electrolytes other than the material being studied must be present in the solution.

Although various techniques for studying the mobility of organic compounds in free solution have been successfully employed, the equipment is usually complex and the methods experimentally cumbersome. Consequently for this discussion only those methods in which the water solution is entrained in a porous medium will be considered. This procedure, which has been designated "zone electrophoresis" by Tiselius to differentiate it from free-space electrophoresis, employs a medium such as paper, cellulose acetate, powdered plastic, or various types of gel to prevent free flow of water and thus minimize density and convection gradients.

Since paper electrophoresis is convenient, has received much attention, and in general can be visualized using the same methods as paper chromatography, it represents the most important single method. In this procedure a paper strip or sheet is wet with buffer, a small sample is applied, and the ends of the sheet are immersed in tanks of buffer containing electrodes. When voltage is applied to the electrodes, electric current flows through the buffer to the paper and through the buffer bridge within the paper. At the electrodes H^+ or OH^- ion is produced and H_2 or O_2 is liberated.

Since current can only be carried by migrating ions, there is a movement of the ions of the buffer through the circuit. All of the current can be carried by either cations or anions. With most buffers some of each type of ion migrates. The difference in the movement of cations and anions and the degree of hydration of each type usually produce a net movement of water within the paper. The immobile anions due to the carboxyl content of paper result in a net movement of water toward the cathode since mobile hydrated cations migrate in a higher percentage of the total. The water movement is termed electroosmosis. The net movement of a particular substance, due to its ionic nature, cannot be determined unless the extent of water movement is known. Since this cannot be predicted in advance it is usually advisable to include a neutral detectable compound on the same sheet or in the solution containing the sample. Since the electroosmosis effect is due to primarily nonmobile charges (usually COOH of the paper) the effect diminishes at lower pH values and higher ionic strength buffer.

The relationship between the observed mobility, U_{eff}, of a charged ion, its inherent mobility, U, and the pK was described by Consden, Gordon, and Martin (*7*).

$$U_{eff} = \frac{UK_d}{(H^+) + K_d} \tag{II.9}$$

since at the pK

$$K_d = (H^+)$$

then at the pH one-half value

$$U_{eff} = \tfrac{1}{2}U \tag{II.10}$$

That is, the pH giving half-maximum mobility is the pH one-half value of the compound.

This determination obviously depends upon carrying out the electrophoresis at a number of different pH values under conditions in which the effects of the ionic background are constant and on the assumption that the paper or other support is either inert or does not selectively adsorb either the ionized or nonionized species.

Unfortunately, buffers having constant ionic strength at a variety of pH values are difficult to obtain and hence some approximation is required.

Since the mobility is a function of the charge per unit mass of the compound being investigated the use of reference standards of known mobility has been employed to determine the equivalent mobility of an unknown compound (*8*). In this procedure a pH at which the compound has maximum mobility is used. The reference standards employed are a neutral dyestuff, a charged dyestuff, and an amino acid of known equivalent weight.

The isoelectric point of amphoteric compounds is readily determined by the electrophoretic method. This is conveniently carried out by plotting mobility at various pH values vs. pH. The point of zero mobility is then obtained from the plot. This simple procedure becomes rather complicated in actual practice. This is due to the effect of pH and ionic strength on electroosmosis. It is therefore usually necessary to incorporate neutral standards as markers which can be used to determine the true zero point of each electrophoresis run.

Thus, Raacke and Li (*9*) showed that carefully controlled electrophoresis in starch gel layers at different pH values could be used to obtain isoelectric points of proteins and peptides. The values were shown to agree with those obtained by free-space electrophoresis. Date obtained with α-corticotropin are presented in Table II.2.

TABLE II.2

ELECTROPHORETIC MOBILITY OF α-CORTICOTROPIN

Buffer	pH	Observed mobility	Electroosmosis[a]	Net mobility
Acetate	4.85	10.6	4.3	6.3
Acetate	5.80	8.3	5.1	3.2
Cocodylate	6.75	6.3	7.2	−0.9
Cocodylate	7.10	5.5	8.6	−3.1
Veronal	7.8	0.5	8.1	−7.6

[a] Two proteins of known mobility were used to correct for water movement rather than a neutral compound.

A variety of equipment types are available for paper electrophoresis. For determination of relative mobilities under different conditions as outlined above it is desirable to conduct the experiments under nearly identical conditions of temperature and electric field. Buffer reservoirs must maintain constant pH. Since zone spreading because of diffusion is an important aspect of electrophoresis, this is minimized by using the highest voltage possible. Equipment for very high voltage short-time separations is available. In this equipment the strip is usually immersed in a cooled water-immiscible liquid. Obviously, the method can only be used with compounds which are not soluble in the coolant.

The voltage and time required for a certain migration vary considerably with the experimental conditions. However, a power supply capable of supplying about 1000 V is usually adequate. Cooling is preferably carried out using a constant-temperature coolant.

The method of handling the developed strip is important and improper handling can cause misleading results. Since the water within the wet paper can move, strips must be handled in such a way that this movement is minimized. Strips should not be hung for drying or exposed to heating or cooling which might preferentially dry one end or the other. Very wet strips should be held horizontally until dry.

In addition to usefulness in determining the ionic nature of an unknown substance, paper electrophoresis is especially effective in separating (on an analytical scale) compounds which may be difficult to separate by conventional paper or thin-layer chromatography. This is especially true for polar compounds which are difficult to characterize by chromatography. Since the net migration is influenced by the nature of the support, the distribution of pore sizes, the functional factors due to the support, and the nature of buffer ions, closely related compounds may give differential migration rates.

D. Polarity

1. Introduction

The term "polarity" has already been briefly discussed. Some relationship between dielectric constant and polarity exists. Practically all organic compounds contain polar and nonpolar functions and the general properties of solubility and adsorbability of molecules of similar size can be loosely correlated to polarity. The behavior of an unknown substance in microchemical tests for solvent extractability and in paper and thin-layer chromatographic systems reflects its overall polarity. Relationships with known compounds can be useful in predicting systems which can be employed for both group separation and fractionation procedures.

2. *Polarity Scale*

As a preliminary guide the correlations in Table II.3 may be used.

TABLE II.3

ARBITRARY POLARITY SCALE

Separation method	Degree of Polarity								
	Highly polar		Medium polarity				Highly nonpolar		
	9	8	7	6	5	4	3	2	1
CCD	Poor		Good		Very good		Good		Poor
Partition chromatography	Poor		Good	Very good			Good		Poor
Ion exchange	Very good			Good			Poor		
Ion exclusion	Very good		Good			Poor			
Adsorption I[a]	Poor		Good			Very good			Good
Adsorption II[a]	Good		Very good				Good		Poor

[a] Adsorption I, nonpolar adsorbents, Adsorption II, polar adsorbents.

An arbitrary classification is as follows (K_1 is the partition coefficient in *n*-butanol/water; K_2 in ether/water; K_3 in hexane/acetonitrile).

Class 9—soluble in water, poorly soluble in methanol, insoluble in isopropanol or acetone; K_1, 0.01.

Class 8—soluble in water, methanol, poorly soluble in isopropanol, insoluble in acetone; K_1, 0.01–0.1.

Class 7—soluble in water, methanol, poorly soluble in butanol, acetone, insoluble in ethyl acetate; K_1, 0.1–0.2; K_2, 0.005–0.001.

Class 6—soluble in water, slightly soluble in ethyl acetate, insoluble in ether; K_1, 0.2–1.0; K_2, 0.02–0.2.

Class 5—poorly soluble in water, soluble in methanol, ethyl acetate, ether; K_1, 0.5–2.0; K_2, 0.1–2.

Class 4—poorly soluble in water, soluble in butanol, ethyl acetate, ether; K_1, 2–10; K_2, 1–10.

Class 3—insoluble in water, soluble in butanol, ethyl acetate, ether, less soluble in benzene, hexane; K_1, 10–100; K_2, 5–50.

Class 2—slightly soluble in butanol, soluble in ether, benzene, hexane; K_2, 40–100; K_3, 0.10–0.1.

Class 1—insoluble in butanol, soluble in ether, hexane; K_2, 2–100; K_3, 0.1–10.

TABLE II.4

PARTITION RATIOS AND POLARITY CLASS OF SIMPLE ORGANIC COMPOUNDS

Compound	Partition ratio with water		
	Isobutanol	Ether	Class
Malonamide	0.086	0.00030	7
Methylamine	0.62	0.023	6
Formamide	0.22	0.0014	6
Acetamide	0.33	0.0025	6
Glycolic acid	0.34	0.028	6
Oxalic acid	0.5	0.12	6
Malonic acid	0.70	0.10	6
Succinic acid	0.96	0.15	6
Lactic acid	0.70	0.09	6
Ethylamine	1.2	0.06	5
Propylamine	3.7	0.29	5
Butyramide	1.5	0.058	5
Dimethylmalonic acid	4.9	1.6	5
Adipic acid	3.5	0.54	5
α-Hydroxyisobutyric acid	1.2	0.26	5
Acetic acid	1.2	0.52	5
Butylamine	9.2	1.1	4
Propionic acid	3.3	1.8	4
Butyric acid	9.4	6.5	4
Chloroacetic acid	2.5	2.9	4
Bromoacetic acid	3.6	4.4	4
Iodoacetic acid	5.9	7.2	4
Pimelic acid	7.3	1.5	4
Hexylamine	83	16	3
Triethylamine	21	5.9	3
Isovaleric acid	20	20	3
α-Bromoisobutyric acid	27	45	3
Phenylacetic acid	27	37	3

The general relationship between groups is illustrated by the following examples:

Glucose—1 g dissolves in 1.1 ml of water, 120 ml of methanol. Soluble in hot glacial acetic acid, pyridine, and aniline. Partition ratio, isobutanol/water, 0.011; ether/water, 0.0000045. Class 8.

Acetamide—1 g dissolves in 0.50 ml of water, 2.0 ml of ethanol, and 6 ml of pyridine. Soluble in chloroform, glycerol, and hot benzene. Partition ratio, isobutanol/water, 0.33; ether/water, 0.0025. Class 6.

TABLE II.5

INFLUENCE OF ADDED GROUPS ON R_f AND DISTRIBUTION COEFFICIENT (G) OF PHENOLS

	R_f and $(G)^a$							
Compound[b]	C_6H_6		$C_2H_4Cl_2$		$(n\text{-}C_4H_9)_2O$		$n\text{-}C_4H_9OH$	
RNO_2	0.55	(1.2)	0.81	(4.2)	—		—	
$ROCH_3$	0.32	(0.47)	0.74	(2.8)	—		—	
RH	0.11	(0.12)	0.48	(0.92)	—		—	
$RCH{=}NOH$	0.03	(0.03)	0.13	(0.15)	0.49	(0.98)	—	
RCOOH	0.00	—	0.06	(0.06)	0.53	(1.1)	—	
ROH	0.01	(0.01)	0.05	(0.05)	0.22	(0.28)	0.85	(5.6)
RNH_2	0.04	(0.04)	0.11	(0.12)	0.05	(0.05)	0.76	(3.1)

[a] Calculated by assuming similar cross sectional area and moisture content for all systems.

[b] R = HO–C_6H_4–

TABLE II.6

DIELECTRIC CONSTANTS OF RELATED COMPOUNDS

Compound	Dielectric constant at 20° or 25°C	Compound	Dielectric constant at 20° or 25°C
CH_3CONH_2	59	$CH_3N(CH_3)_2$	2.44
CH_3CN	37	CH_3Cl	12.6
CH_3OH	32.6	CH_2Cl_2	9.08
CH_3CHO	21.8	$CHCl_3$	4.81
CH_3COCH_3	20.7	CCl_4	2.24
CH_3COCl	16.0	CS_2	2.64
CH_3COOCH_3	6.68	CH_4	1.70
CH_3COOH	6.15 (dimeric)	CH_3CH_2OH	24.3
CH_3NH_2	11.4	$CH_3CH_2NO_2$	28.0
CH_3NHCH_3	5.3	$CH_3CH_2CH_2CH_2OH$	17.1

Caffeine—1 g dissolves in 46 ml of water, 66 ml of ethanol, 100 ml of benzene, and 530 ml of ether (chloroform, 5.5 ml). Partition ratio, isobutanol /water, 1.2; ether/water, 0.06. Class 5.

Benzoic acid—1 g dissolves in 300 ml of water, 2.3 ml of ethanol, 3 ml of ether, 10 ml of benzene, and 30 ml of carbon tetrachloride. Partition ratio, isobutanol/water, 50; ether/water, 78. Class 3.

The use of the polarity scale is at best only an arbitrary guide but it can be useful in a general way. Collected information with precise solubility and partition ratios determined under identical conditions is not available. However, the publications of Collander (*10*) and Sandell (*11*) are valuable in showing the effects of individual substituents on the total polarity. Table II.4 includes data on partition ratio and indicated polarity class for a number of organic compounds.

It can be seen that a shift of one class is caused by a change of two to three methylene groups. A carboxyl may cause a shift of one or two classes, whereas $CONH_2$ causes a shift of two classes.

Aromatic compounds were studied extensively by Sandell (*11*) using paper chromatography. For disubstituted compounds there is sometimes a substantial difference in R_f, depending upon the position of the substituents. Table II.5 illustrates the effect of a single added substituent in the para position of phenol.

A further indication of the relative polarity can be obtained by a comparison of the dielectric constant of the added group. Related compounds are tabulated in Table II.6.

E. Molecular Size

1. Introduction

No information regarding molecular size is obtainable from the above tests. Since most chromatographic processes require pentration of a solid matrix by the materials being separated, some information on molecular size may be useful in choosing the best procedure. Since the tests of interest are those which can be carried out on crude materials many molecular size determinations cannot be used. Molecular size is not as important in separation processes as diffusion rate, which is influenced not only by size but also by molecular shape. The latter may be influenced by environmental factors such as hydrogen bonding solutes and the presence of ionizable or chelating groups. For ionized substances the net mass per charge may be derivable from electrophoretic measurements and the total number of charges per molecule by one of the partition methods outlined below. These methods yield an approximate molecular weight.

2. Dialysis

For neutral materials diffusion rates can be determined if an adequate assay method is available. Craig has studied factors influencing permeability through membranes (dialysis) in detail. These studies are summarized in two review articles (*12, 13*). Dialysis is frequently used as a group separation method to separate "large" from small materials. As a process step, ultrafiltration, a nonequilibrium process, is preferred to dialysis. With cellophane membranes of the usually available types, nonfilterable substances would generally be of 15,000 m. wt. or larger in size. Certain membranes may exclude smaller substances.

3. Gel Permeation

The availability of gel matrixes of varying density has promoted the use of these materials for molecular size determination. The theory of this method will be discussed later. These substances can be used in spread layer plates. Determination of mobility, usually simultaneously with known substances, can be used to approximate molecular size. For R_f values in the region of about 0.2–0.8, the log of the molecular size is approximately linear with R_f.

$$\log \text{m.wt.} = xR_f + c \tag{II.11}$$

If standards are used to determine the slope, the molecular weight of the unknown substance can be approximated.

F. Functional Group Determinations

1. Introduction and Theory

Frequently some idea of the number of ionizable functions per molecule or of some other possible replicating group may be obtained by partition methods, either chromatography or solvent extraction.

The general concept that the distribution properties of an organic molecule represent the sum of the structural components was introduced by Martin (*14*). Complete agreement with the theory in its simplest form would result in identical distribution properties and hence inseparability of structural isomers, such as leucine and isoleucine and the like. Since it is known that many such compounds and even, in certain cases, *d*, *l* isomers can in fact be separated, the general applicability of the theory has been questioned. On the other hand, many examples have accumulated which provide a broad base for general acceptability. Certainly in considering the properties of an unknown substance application of the theory can yield useful information. Difficulties arise when attempts are made to use the theory in its strictest sense and under conditions in which all experimental parameters cannot be controlled with sufficient accuracy.

It has already been shown (Chapter I) that the R_f of a compound in a partition chromatographic system is related to its partition ratio, K, between the two phases of the system employed.

For ideal solutions of substance A, K is related to the free energy, Δu_A, required to transport 1 mole from one phase to the other

$$\ln K = \frac{\Delta u_A}{RT} \tag{II.12}$$

Martin proposed that the addition of a group x to substance A would change the partition ratio by a factor which depended only upon the nature of x and the two phases but not on A itself. Thus if A is substituted by nx groups, my groups, etc.

$$RT \ln K = \Delta u_A + n\Delta u_x + m\Delta u_y + \text{etc.} \quad \text{(II.13)}$$

from Eq. (I.6) it can be shown that

$$K = \frac{V_m}{V_s}\left(\frac{1}{R_f} - 1\right)$$

and

$$RT \ln \frac{V_m}{V_s}\left(\frac{1}{R_f} - 1\right) = \Delta u_A + n\Delta u_x + m\Delta u_y + \ldots \quad \text{(II.14)}$$

2. *R_m and ΔR_m*

The term, R_m, was introduced by Bate-Smith and Westall (*15*).

$$R_m = \log\left(\frac{1}{R_f} - 1\right) \quad \text{(II.15)}$$

These authors showed that the relationship applied to a number of related compounds.

One of the important conclusions from the theory is that the R_m increment (ΔR_m) for a given substituent should be a constant, irrespective of the remainder of the molecule (assuming that other interactions are absent). A review of the studies conducted for several series of compounds and in several systems has been published (*16*).

Lederer and Lederer (*17*) studied $\Delta R_m(CH_2)$ for several homologous series and concluded that $\Delta R_m(CH_2)$ was in fact constant. Recently Howe (*18*) concluded that for a series of compounds—carboxylic acids, dicarboxylic acids, and amino carboxylic acids—a plot of R_m vs. the number of CH_2 groups is linear or nearly so, but that $\Delta R_m(CH_2)$ was not constant in the different series since the slopes were not parallel. Green and Marcinkiewicz (*19*, *20*) have described a chromatographic system which minimizes experimental variation. Using aromatic phenols, these workers have determined ΔR_m for a number of functional groups and indeed have also calculated atomic ΔR_m values for these compounds. Certain causative effects were observed and the magnitude of these effects may depend on the system employed. Thus, although the Martin relationship has been established, it is necessary to apply corrections to calculate the R_f of an unknown compound from ΔR_m values.

A. AMIDE GROUPS IN PEPTIDES. Certain correlations in R_m or K are possible, however. Pardee (*21*) applied the theory to the calculations of R_f values of simple peptides. By assuming that the contributions to

$$RT \ln \frac{V_m}{V_s} \left(\frac{1}{R_f} - 1 \right)$$

could be divided into the components of the peptide and that similar components in the free amino acids were additive, the general expression

$$RT \ln \left(\frac{1}{R_f} - 1 \right)_{\text{peptide}} = (n - 1)A + B + RT \ln \left(\frac{1}{R_f} - 1 \right)_{\text{amino acids}} \qquad \text{(II.16)}$$

was obtained. A is a constant including the contributions of peptide, amino and carboxyl groups; n is the number of amino acid residues; and B is a term introduced to account for the difference between the terminal groups of a peptide and the corresponding amino acid. These would obviously be dependent on the system employed. According to this theory, the R_f values of peptides containing the same amino acids would be identical regardless of structure. Thus glycylvaline would be expected to have the same R_f as valylglycine, glycylvalylalanine the same as valylglycylalanine, etc. These predictions were confirmed by the data from the literature cited by Pardee.

Later Moore and Baker (*22*) reported a study of the R_f values of 88 synthetic peptides and 29 amino acids in 15 solvent systems, the constants, A and B, above were determined from the R_f values of 33 dipeptides and 11 tripeptides in 9 solvent systems. Most of the peptides employed contained glycyl or alanyl residues. The R_f values of these peptides agreed with the calculated values within ± 0.05 for the majority of the compounds. A large exception was noted for L-cystinyl-L-cystine. Occasionally peptides containing sarcosine, glutamic acid, aspartic acid, lysine, or phenylalanine had R_f values differing from the calculated by a value of greater than 0.1. Some difference in the systems used was observed. Very few quantitative data on peptides of longer chain length (especially containing six or more amino acids) are available but it would be expected that increased intermolecular interaction would bring about a greater divergence of the calculated and observed values.

For the study of peptide sequences the theory is of little value. Certain special applications may be useful, however. Thus, cyclic peptides would have an R_f value quite different from that calculated for the linear peptide. An inseparable mixture of two small peptides would have a different R_f than a single peptide containing the same amino acids.

B. OLIGOSACCHARIDES AND OTHER COMPOUNDS. The correlation of R_m values with the number of hexose units in oligosaccharides (*23*, *24*) and with the number of CH_2 units in *N*-*n*-alkyl tritylamines (*25*) have been reported and point to the broad application of the R_m concept. One of the most extensive studies of this type showed that ΔR_m values for hydroxyl, carboxyl, acetoxy, and methyl groups on the steroid nucleus were constant through a series of seven different solvent systems needed to chromatograph the variety of compounds tested (*26*). With this system it is, therefore, possible to determine the number and type of functional groups present in a steroid compound.

3. *Acidic and Basic Groups*

Frequently, some idea of the number of ionizable functions per molecule or of some other possible replicating group may be obtained by partition methods, either chromatography or direct partition.

The concept of contributions to R_m from specific functional groups has interesting possibilities for the determination of the number of reactive functional groups in the molecule. For chromatography in two solvent systems having the same cross sectional area of mobile and stationary phase

$$ZR_{m,1} = \Delta u_{A,1} + n\Delta u_{x,1} + m\Delta u_{y,1} + \ldots$$

$$ZR_{m,2} = \Delta u_{A,2} + n\Delta u_{x,2} + m\Delta u_{y,2} + \ldots$$

The difference in R_m is

$$Z(R_{m,1} - R_{m,2}) = (\Delta u_{A,1} - \Delta u_{A,2}) + n(\Delta u_{x,1} - \Delta u_{x,2}) + m(\Delta u_{y,1} - \Delta u_{y,2}) + \ldots \qquad \text{(II.17)}$$

in which Z is a constant containing solvent-phase ratios, Δu_A expresses the contribution of the nonreactive portion of the molecule and Δu_x and Δu_y represent contributions of the reactive groups. If two identical solvent systems are employed, the A terms disappear and the observed difference is due to the functional groups. If such groups are ionizable and a strongly polar solvent system is used then the reactive functions indicate a difference in the ionizable portion of the molecule. Experimental observations would appear to confirm the general applicability of this approach. Using *n*-propanol–water (7:3) containing 2 N NH_4OH or SO_2, a relatively constant R_m difference of about 0.50 was observed *per* carboxyl group for a large series of mono- and dicarboxylic acids (*27*). The R_m difference for tricarboxylic acids appeared to be lower. Very strong acids also had a lower R_m. In another study the same propanol solvent containing 2 N methylamine or 2 N isopropylamine was utilized (*28*). Since isopropylamine is present at nearly 5% of the total solvent concentration in this case, the term $(\Delta u_{A,1} - \Delta u_{A,2})$ is probably not negligible. For compounds with pK values greater than about 10^{-6} $(\Delta u_{A,1} - \Delta u_{A,2})$ appears to contribute 0.08 R_m and $(\Delta u_{x,1} - \Delta u_{x,2})$ about 0.13 per carboxyl. Thus, $R_{m,1} - R_{m,2}$ for monocarboxylic acids, and some inorganic acids was observed to be 0.21, for dicarboyxlic acids, 0.34; for tricarboxylic acids, 0.48.

For amino acids, the differences between acidic and basic systems are more pronounced. However, dicarboxylic monoamino acids have characteristic differences from monocarboxylic monoamino acids.

Although chromatography can be rapidly carried out with a number of different solvent systems and the R_f determined accurately, the influence of the support (paper or thin layer) on the system may not be constant. In addition any solvent demixing during the chromatographic run causes changes in composition and relative location. Direct determination of the partition ratio in a solvent system eliminates uncertainty due to these effects. From Eq. (II.13) it is seen that for two closely related solvents in which the backbone effect is similar, the contribution of similar functional groups, $n\Delta u_x$, should be a function of the number and character of the functional group. Using 2-butanol and 0.1 *N* aqueous solutions of strong monobasic acids, Carpenter *et al.* (*29*) observed that the partition ratio for a number of amino acids and peptides could be correlated with the partition ratio of the acid itself. Using a log–log plot of these partition ratios, a straight line having a slope of n was obtained, in which n is the number of basic groups in the molecule.

The data presented in Table II.7 are amenable to analysis in terms of increments and a determination of n is possible from this procedure.

Examination of the above data reveals remarkably close correlation for n, the number of basic groups, in the compounds tested for the simpler compounds. Since the partition ratios were determined by total ninhydrin color of the acid hydrolysates of each phase, impurities contributing to the total ninydrin value would influence the result drastically. Thus, the results for bacitracin A and insulin may indicate slight inhomogeneity in the compounds tested rather than nonconformity of the method.

G. Summary

Since the preparation of some natural compounds of biochemical interest in gram quantities may require processing of extremely large quantities of starting materials, the group separation step employed to obtain a crude concentrate can be of critical importance. The advantages and disadvantages of individual methods will be discussed later.

The main purpose of the preliminary testing discussed here is to provide an indication of possible methods or approaches which can be used for a particular problem. For a multistep process each successive step should be capable of removing impurities which may have been included in the previous step. Since fractionation steps are generally difficult in manipulation, expensive in time and assay effort, and are not readily applied to large quantities of material, the group steps should be designed to pare away dissimilar compounds. This can be done by locating the desired compound in the above

TABLE II.7

PARTITION RATIOS OF COMPOUNDS IN 0.1 *N* ACID SYSTEMS[a]

Compound	$HOCH_2CH_2SO_3H$ log *K*	HBr log *K*	Δlog[b]	HNO_3 log *K*	Δlog[b]	HCl[c] log *K*	HNO_3[c] log *K*	Δlog[c]
Acid	−0.58	−0.40	0.18	−0.32	0.26	−0.40	−0.123	0.28
Ammonia	−1.17	−0.94	0.23	−0.86	0.31	−0.86	−0.70	0.16
Glycine	−1.09	−0.89	0.20	−0.82	0.27	—	—	—
Alanine	−1.07	−0.82	0.25	−0.74	0.33	−0.75	−0.56	0.19
α-Aminobutyric acid	−0.89	−0.68	0.21	−0.61	0.28	—	—	—
Norvaline	−0.65	−0.42	0.23	−0.34	0.31	—	—	—
Avg.	—	—	0.21	—	0.30	—	—	0.18
Arginine	−1.64	−1.28	0.36	−1.16	0.48	—	—	—
Lysine	−1.82	−1.45	0.37	−1.34	0.48	−1.20	−0.92	0.28
Ornithine	−1.82	−1.49	0.33	−1.36	0.46	−1.21	−0.92	0.29
Histidine	−1.89	−1.56	0.33	−1.51	0.48	−1.18	−0.88	0.30
Arginylleucine	−0.93	−0.53	0.40	−0.40	0.53	—	—	—
Arginylglutamic acid	−1.44	−1.10	0.34	−0.96	0.48	—	—	—
Glutathione disulfide	−1.62	−1.29	0.33	−1.19	0.43	—	—	—
Avg.	—	—	0.35	—	0.48	—	—	—
Histidylhistidine	—	−1.70	—	−1.40	—	−1.54	−1.15	0.39
Gramicidin S–A	0.98	1.17	0.19	1.50	0.52	—	—	—
Bacitracin	−0.36	0.19	0.55	0.38	0.74	—	—	—
Insulin	−1.54	−0.56	0.98	−0.25	1.29	—	—	—
Cystine	—	—	—	—	—	−1.21	−0.92	0.29
$(\Delta u_{A,1}-\Delta u_{A,2})$	—	—	0.07	—	0.10	—	—	0.09
$(\Delta u_{x,1}-\Delta u_{x,2})$	—	—	0.14	—	0.19	—	—	0.10

[a] See ref. (*29*).

[b] 2-Butanol, Δ log to $HOCH_2CH_2SO_3H$ system.

[c] 2-Butanol + 3% methanol, Δ' log between system.

polarity scale and eliminating compounds of other polarity by some procedure. The resultant crude product, now somewhat uniform in polarity, is more amenable to fractionation.

These same principles apply to isolation of a substance from a synthetic reaction mixture. Frequently this type of operation may be carried out in routine fashion by the chemist. The removal of unreacted acid from an esterification reaction by extraction of a solvent solution with a base is an example. For complex reaction mixtures a group separation method may not be practical. Some sort of fractionation step may be required to isolate the desired component. If the polarity properties of the starting compounds are known, the above guide may be useful in choosing a fractionation system and in predicting the general location of the products. For this purpose the relative dielectric constants of various functional groups may be useful.

Consider the reaction of an ester with a primary amine to form an amide, $RCOOCH_3 \rightarrow RCONHCH_3$. The reaction product would be considerably more polar than the starting ester. Examination of the reaction mixture at various times or after certain types of treatment by a thin-layer or paper strip chromatographic system in which the mobility of the parent ester is known would yield information concerning the course of the reaction. Multiple products from a reaction may be revealed in a similar fashion, e.g., $R_2CHCHClR' + B^+ \rightarrow R_2CHCHBR' + R_2C = CHR'$. The large difference in polarity of the substitution product compared with the elimination product would usually be readily detectable in a chromatographic system in which the parent had reasonable mobility. Thus these small-scale tests and chromatographic methods can provide useful insight into the course of a reaction.

REFERENCES

1. Rothrock, J. W., Goegelman, R. T., and Wolf, F. J., *Antibiot. Ann.* p. 796 (1958–1959).
2. Betina, V., *Nature* **182**, 796–797 (1958).
3. McKinley, W. P., in "Analytical Methods for Pesticides, Plant Growth Regulators, and Food Additives" (G. Zweig, ed.), Vol. 1, p. 243. Academic Press, New York, 1963.
4. Golumbic, C., Orchin, M., and Weller, S., *J. Am. Chem. Soc.* **71**, 2624–2627 (1949).
5. Badgett, C. C., Eisner, A., and Walens, H. A., *J. Am. Chem. Soc.* **74**, 4096–4098 (1952).
6. Waksmundski, A., and Soczewinski, E., *J. Chromatog.* **3**, 252–255 (1960).
7. Consden, R., Gordon, A. H., and Martin, A. J. P., *Biochem. J.* **40**, 33–41 (1946).
8. Werum, L. N., Gordon, H. T., and Thornburg, W., *J. Chromatog.* **3**, 125–145 (1960).
9. Raacke, I. D., and Li, C. H. J., *J. Biol. Chem.* **215**, 277–285 (1955).
10. Collander, R., *Acta Chem. Scand.* **4**, 1085–1098 (1950); *Acta Chem. Scand.* **5**, 774–780 (1951).
11. Sandell, K. B., *Monatsh. Chem.* **89**, 36–53 (1958).
12. Craig, L. C., and King, T. P., *Methods Biochem. Analy.* **10**, 175–199 (1962).
13. Craig, L. C., *Science* **144**, 1093–1099 (1964).
14. Martin, A. J. P., *Biochem. Soc. Symp.* (*Cambridge, Engl.*) **3**, 4–20 (1950).

15. Bate-Smith, E. C., and Westall, R. G., *Biochim. Biophys. Acta* **4**, 427–440 (1950).
16. Bush, I. E., *Methods Biochem. Analy.* **13**, 357–438 (1965).
17. Lederer, E., and Lederer, M., "Chromatography," 2nd ed. Elsevier, Amsterdam, 1957.
18. Howe, J. R., *J. Chromatog.* **3**, 389–405 (1960).
19. Green, J. S., and Marcinkiewicz, S., *J. Chromatog.* **10**, 35–41 (1963).
20. Green, J. S., and Marcinkiewicz, S., *Chromatog. Rev.* **5**, 58 (1963).
21. Pardee, A. B., *J. Biol. Chem.* **190**, 757–762 (1951).
22. Moore, T. B., and Baker, C. G., Jr., *J. Chromatog.* **1**, 513–520 (1958).
23. French, D., and Wild, G. M., *J. Am. Chem. Soc.* **75**, 2612–2616 (1953).
24. Thoma, J. A., and French, D., *Anal. Chem.* **29**, 1645–1648 (1957).
25. Boyce, C. B. C., and Milborrow, B. V., *Nature* **208**, 537–540 (1965).
26. Kabasakalian, P., and Basch, A., *Anal. Chem.* **32**, 458–461 (1960).
27. Reichl, R., *Mikrochim. Acta* pp. 955–965 (1956).
28. Borst-Pauwels, G. W. F. H., and DeMots, A., *J. Chromatog.* **15**, 361–370 (1964).
29. Carpenter, F. H., McGregor, W. H., and Close, J. A., *J. Am. Chem. Soc.* **81**, 849–855 (1959).

III *Solvent Extraction*

The following discussion will consider liquid–liquid extraction procedures which rely on the ability to transfer the desired substance from one immiscible liquid to another or the potential removal of impurities by extraction of a liquid solution with an immiscible liquid. The operation consists of two steps. The liquids are intimately mixed until the solutes have been distributed between the liquids and the phases are separated.

The immiscible liquids may be any combination of two or more solvents. Many useful solvent "pairs" contain four or more components. In general solutes have some solubility in both phases of the immiscible liquids.

I. PARTITION RATIO

A. Definition

The ratio of the concentration of the solute in each of the liquids can be determined by analysis and is the partition ratio, K

$$K = C_u / C_l \tag{III.1}$$

where u indicates upper phase and l, lower phase.

B. Chemical Potential

1. Activity Coefficient

Since, at equilibrium the chemical potential or activity of the solute in each of the phases is equal, the partition ratio is the ratio of the solubility of the solute in each phase. Since the chemical potential, u_A, of solute A at a concentration, α, expressed in mole fractions, with reference to some standard state u° is $u_A = u^\circ{}_A + RT \ln \alpha$, at equilibrium between two solvents it is $u_{A,1} = u^\circ_{A,1} + RT \ln \alpha_1 = u^\circ_{A,2} + RT \ln \alpha_2$ and

$$\ln \frac{\alpha_1}{\alpha_2} = \ln K_1 = \frac{u^\circ_{A,1} - u^\circ_{A,2}}{RT} = \frac{\Delta u_A}{RT} \tag{III.2}$$

in which Δu_A is the free energy required to transport 1 mole of solute A from phase 1 to phase 2. This relation applies to each "species" of solute.

Thus for compounds containing a COOH group, the overall partition ratio would include individual contributions due to the dimeric species, any monomeric species, and any hydrogen-bonded complex with a solvent constituent. The net free energy is influenced by solvent–solvent, solute–solute, and solvent–solute interactions. For compounds containing polar groups and polar solvents, these interactions can be substantial. In a study of the free energy relationship of solutions, Pierotti *et al.* (*1*) observed that the logarithms of activity coefficients for members of homologous series fall into simple correlation patterns. These patterns were interpreted empirically based on molecular interactions. By assuming that the nature of the interactions is qualitatively the same for the same functionality in both solute and solvent, the terms of the following relationship were studied:

$$\log \gamma^\circ = Z_{1,2} B_2 \frac{n_1}{n_2} + \frac{C_1}{n_1} + D(n_1 - n_2)^2 + \frac{F_2}{n_2} \quad \text{(III.3)}$$

in which $\log \gamma^\circ$ is the activity coefficient of component R_1X_1 at infinite dilution in component R_2X_2; n_1 and n_2 are carbon numbers in R_1 and R_2, respectively: $Z_{1,2}$ is a coefficient which depends on the nature of solute and solvent functional groups; B_2 is a coefficient which depends only on the nature of the solvent functional group, X_2; C_1 is the same for solute functional group Z_1; D is an exponential coefficient independent of both X_1 and X_2; and F_2 is a second coefficient which essentially depends on solvent functional group, X_2.

Pierotti calculated the γ values for several classes of compounds in simple solvents and found a remarkable correlation within a homologous series. However, an *exact* prediction of the partition ratio of a particular solute is not possible from these data. Predictions of group type separations were shown to be possible. If data are available for one compound of a homologous series, extrapolation to other compounds can be attempted.

2. *Relationship between Different Solvent Systems*

For systems obeying Raoult's law the following relationship would apply for two-solvent systems and one solute:

$$\ln K_1 = \ln K_2 + \frac{\Delta u_{A,1} - \Delta u_{A,2}}{RT} \quad \text{(III.4)}$$

This relationship expresses a useful concept even though most systems deviate from Raoult's law. The presence in expression (III.3.) of the exponential term D in the logarithm of the activity coefficient indicates that even for the same solute in two homologous solvent systems the partition ratio would be difficult to predict in the absence of experimental data. Collander (*2, 3*) studied a variety of organic compounds and observed that for a homologous series, the general expression

$$\log K_1 = a \log K_2 + b \tag{III.5}$$

appeared to correlate with the experimental findings. Thus, apparently the similar interacting effects allow a grouping of the exponential coefficient in the a constant. For a series of alcohol/water systems (butanol, isobutanol, isopentanol, and octanol) the value of a did not exceed 1.25. For several non-ionized compounds the constant b did not differ markedly. Thus, Eq. (III.5) is useful in extrapolating from system to system for compounds of the same functionality.

For a series of relatively nonpolar compounds containing a single functional group (alcohols, esters, acids, aldehydes, and ketones) a gas chromatographic analysis was used to determine K in several three-component solvent

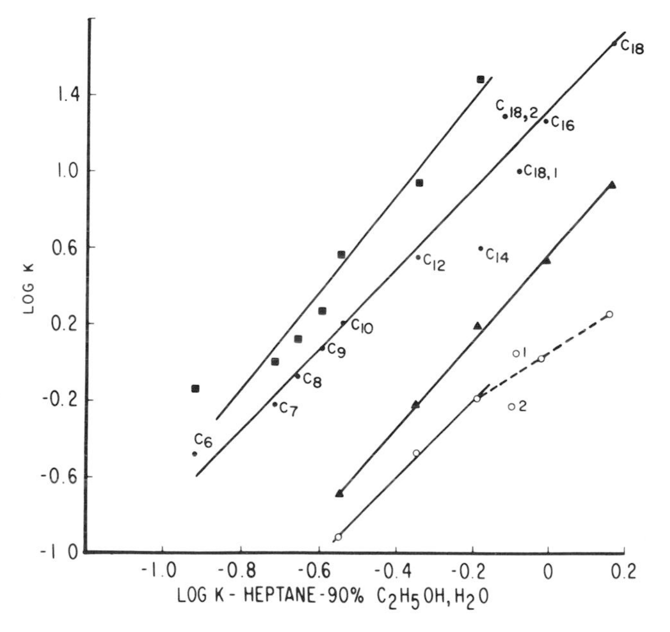

FIG. III.1. Partition ratio of fatty acids. Correlation of log K in ternary systems. Straight chain saturated fatty acids except $C_{18,1}$-oleic acid and $C_{18,2}$-linoleic acid.

	Ordinate	a	b
$\log K = a \log K_1 + b$	● isooctane–80% CH_3COCH_3	2.10	1.35
	○ isooctane–90% $HCON(CH_3)_2$	2.05	0.22
	▲ isooctane–90% $(CH_3)_2SO_2$	2.25	0.58
	■ benzene–80% CH_3OH	2.55	1.92

systems (*4*). Correlation plots of these data indicate a general agreement with expression (III.5). The data for straight-chain acids are contained in Fig. III.1 and those for alcohols in Fig. III.2. High or low values of K are difficult to determine accurately and divergence from straight lines at those points below a K of about 0.1 is not surprising. The values for oleic and linoleic acid are also included and illustrate differences between homologous series.

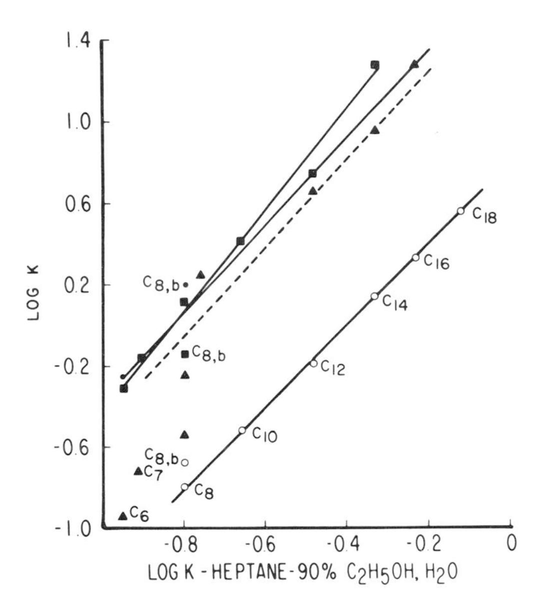

FIG. III.2. Partition ratio of primary alcohols. Correlation of log K in ternary systems. Straight chain saturated primary alcohols except $C_{8,b}$-2-ethyl-1-hexanol.

	Ordinate	a	b
$\log K = a \log K_1 + b$	● isooctane–80% CH_3COCH_3	2.65	2.02
	○ isooctane–90% $HCON(CH_3)_2$	2.00	0.8
	▲ isooctane–90% $(CH_3)_2SO_2$	2.2	1.69
	■ benzene–80% CH_3OH	2.5	2.09

Recently, in a study of the paper chromatographic behavior of carbohydrates in polar systems, Perisho, Rohrer, and Thoma (*5*), indicated that a good correlation of the experimental data is obtained using the assumption that $\Delta u_1/\Delta u_2$, the free energy ratio, is constant rather than the ratio of the partition ratios. In the systems reported water-miscible solvents were used with varying proportions of water and the paper chromatography was conducted in the absence of an immiscible phase.

The effect on K of adding a third solvent miscible with one of the two phases but which does not influence the composition of the other phase is as follows:

for solvent 1

$$\ln K_1 = \frac{\Delta u_{A,1}}{RT}$$

for solvent 2

$$\ln K_2 = \frac{\Delta u_{A,2}}{RT}$$

and for x mole fractions of solvent 1 and y mole fractions of solvent 2

$$x \ln K_1 + y \ln K_2 = \frac{x\,\Delta u_{A,1} + y\,\Delta u_{A,2}}{RT} = \ln K \tag{III.6}$$

The validity of this relationship for changes in composition of the polar phase (isopropyl ether–dimethyl sulfoxide, water) and of the nonpolar phase (benezene, cyclohexane formamide and cyclohexane, chloroform–formamide) was tested by a paper chromatographic technique and confirmed by CCD (*6*).

The R_m values reported (*7*) for 2,5-dihydroxybenzofuran are as given in Table III.1.

TABLE III.1

R_m VALUES FOR 2,5-DIHYDROXYBENZOFURAN[a]

Dimethyl sulfoxide (vol. %)	R_m
40	−0.59
52	−0.17
59	0.30
79	0.90
88.5	1.35
95.5	1.55

[a] Stationary phase, dimethyl sulfoxide-water; mobile phase, isopropyl ether.

This relationship can be useful in modifying the extractive properties of a known system or in extrapolating to a desired partition ratio from limited data. The values from Table III.1 are plotted in Fig. III.3 and indicate clearly that within the region explored the extrapolation would be satisfactory.

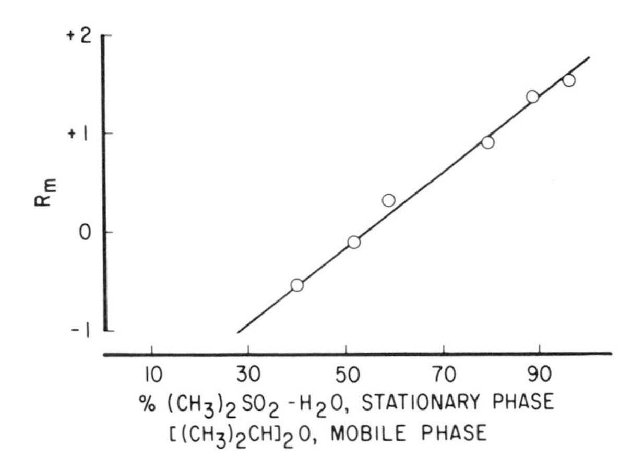

FIG. III.3. Relationship of solvent composition and R_m for 2,5-dihydroxybenzoforan.

A. USE OF TERNARY DIAGRAMS. Usually three-component systems are such that the composition of both phases changes as the ratio of solvents is varied. Under these circumstances no straightforward relationship of the partition ratio and system composition can be formulated. A ternary diagram for the system under consideration can be useful in considering the probable effect of changing ratios of the components. Such diagrams are available for many common solvents or can be readily constructed using gas–liquid chromatography for the analysis of phase compositions. A hypothetical diagram for three solvents, A, B, and C, is given in Fig. III.4. A mixture prepared with the composition, M (21% A, 64.5% B, and 14.5% C) separates into two phases having the compositions x (94.5% A, 2% B, and 3.5% C) and y (15.5% A, 69% B, and 10% C) as represented by the tie line x,y. The relative amount of phases x and y is represented by the distances My and Mx, respectively. As the tie lines approach P, the plait point, the difference in composition between phases x and y becomes less. Although the partition ratio becomes more nearly 1 and hence approaches the ideal point for CCD separation, this may have several disadvantages. Systems near the plait point are easily disturbed by temperature and solute concentration. The density difference is slight and the distribution ratio of all solutes approaches 1, with consequent loss of separating power.

However, there are many cases when a three-component system is useful for both group and fractionation separation. A procedure for the recovery of novobiocin, an acidic nonpolar antibiotic, by solvent extraction has been described (*8*). Novobiocin acid is readily extracted from dilute aqueous solution by common solvents, such as amyl acetate. Difficulty is encountered in attempting a group separation of nonpolar acids by extraction into alkaline

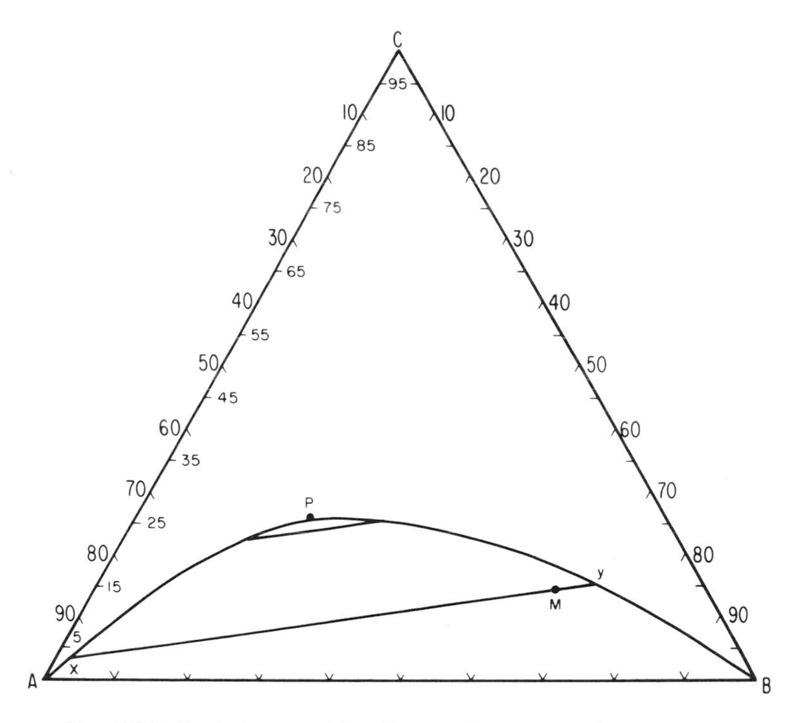

FIG. III.4. Typical composition diagram for ternary solvent system.

water because novobiocin is a weak acid and is unstable at the high pH required. However, the addition of methanol to the alkaline system increases the distribution ratio into the polar phase and permits the use of a lower pH for the separation.

B. SYSTEMS WITH INCREASING POLARITY. For use in fractionation, King and Craig (*9*) have suggested the following series of mixed solvents for studying compounds of increasing polarity:

(a) Hexane or cyclohexane, ethanol, water
(b) Benzene, methanol, water
(c) Chloroform, methanol, water
(d) Ethyl acetate, water
(e) *n*-Butanol or *sec*-butanol, water
(f) Phenol, water

Fractionation of the salts of weak acids or bases requires the use of an adequately buffered system because of the large effect of pH on the partition ratio (Eq. II.4). Salts of strong acids or bases may be rendered less polar and more solvent seeking by addition of "carriers" to achieve a good partition ratio of the ionized substance at or near neutral pH. The effect of the partition ratio of the carrier has been described by Carpenter *et al.* (*10*). In this study the distribution of a base was correlated with the partition ratio of the carrier acid. As the solvent-seeking properties of the acid increase a corresponding increase in the solvent extractability of the salts is obtained.

Materials useful for increasing the partition ratio of polar substances usually have a large nonpolar hydrocarbon or aromatic portion which tends to solubilize the desired compound in the less polar phase and insolubilize it in the polar phase. For acids, amines such as triethylamine, tri-*n*-butylamine, long-chain fatty amines, and their quaternary derivatives may be used. For bases, acids such as 2-ethylhexoic acid, di-2-ethylhexylphosphoric acid, toluenesulfonic acid, and highly chlorinated phenols may be useful. High concentrations of buffers combined with polar solvents have been used in the fractionation of nucleic acids (*11*).

C. Methods of Determining the Partition Ratio

1. Paper Chromatography

Paper strip procedures can be used for an approximation of *K*. This procedure is a convenient way of studying a number of systems rapidly to select certain combinations for further study. Since the procedure does not require a precise quantitative determination of the substance being investigated it may be the only method applicable for crude materials for which no precise assay method is available.

2. Equilibration between Two Phases

For pure material the substance can be partitioned between known volumes of each phase and the dry weight contained in an aliquot of each phase determined. For crude or impure substances an analytical procedure specific for the desired substance is required. If impurities are known to interfere in the analytical procedure a multiple extraction may be useful. An accurate determination of *K* will generally require a precise analysis. The use of gas–liquid chromatography as the analytical method has been mentioned. A multiple extraction procedure is as follows:

Using three extraction vessels (15-ml centrifuge tubes are convenient) all of the material is charged into the first vessel. Equal amounts of stationary phase are added to all three vessels. After equilibration the mobile phase from the first vessel is separated and added to the second vessel and fresh solvent is added to the first vessel. After an additional set of transfers of this type, a three-stage countercurrent extraction has been carried out and six phases have been produced for analysis, upper phases 0, 1, and 2 and lower phases 0, 1, and 2. For a substance having a partition ratio of 1 the relative amounts in each tube are predicted by the numerical terms of the binomial expansion $(p+q)^2$; i.e., 1,2,1. [If four vessels were used, three transfers of the mobile phase present in the orginal vessel would be carried out and the numerical terms 1,3,3,1 of the expansion $(p+q)^3$ would describe the distribution.] The partition ratio can be approximated by a single analysis of any phase. If all six phases are analyzed at least 22 calculations of K can be made. Data which might be obtained from such a run in which 500 μg of material was charged to the first tube are as follows:

Analysis of CCD Using Three Tubes

	Tube 0 [amt. (%)]	Tube 1 [amt. (%)]	Tube 2 [amt. (%)]	Total
Upper	55 (16.9)	153 (47.0)	118 (36.4)	326
Lower	32 (17.7)	78 (43.1)	71 (39.2)	181
Total	87 (17.1)	231 (45.5)	189 (37.2)	507
Fraction expected	$(1-p)^2$	$2p(1-p)$	p^2	—
		p found		
Upper	0.59	0.624	0.6	
Lower	0.58	0.69	0.624	
Total	0.585	0.65	0.61	
Mean p = 0.615		*Range* p = 0.69–0.58		
K = 1.60		K = 2.26–1.38		

The ratio of concentrations in any one tube or for the total in upper and lower phases gives the following values for K:

K Found			
1.72	1.96	1.66	1.80

Similarly the ratios of the quantities found in each phase or tube and that in the adjacent tube can be used.

	Ratios between tubes, K found		
	Tubes 1/0 ($2K$)	Tubes 2/1 ($\frac{1}{2}K$)	Tubes 2/0 (K^2)
Upper	1.39	1.54	1.47
Lower	1.22	1.82	1.48
Total	1.32	1.63	1.47
Mean	1.47		

The above example serves to illustrate the difficulties attendant in determining K accurately since the same data give values as high as 2.26 and as low as 1.22. If inaccuracy is due in part to the lack of complete transfer of the separated phase then the second procedure, which is not influenced by this, would be the more nearly correct.

3. Other Procedures

A third method for the determination of K is the use of a small partition column. For this determination it is necessary to know the volume of stationary phase within the column. The column effluent is monitored to establish the retention volume, V_e, which then permits the determination of K from Eq. (I.11).

A CCD run provides the most precise determination of K. Using radioactive compounds which can be analyzed precisely, K can be determined within ± 0.01 (*12*).

II. APPLICATION OF SOLVENT EXTRACTION FOR GROUP SEPARATION

Solvent extraction may be applied at any stage of a purification procedure but is usually most useful at the beginning of an isolation procedure. High yields can be obtained if the substance is stable and recovery from the solvent is not difficult. Although in the laboratory large amounts of solvent and several extractions may be used to obtain high yields, in manufacturing processes the use of more than two successive extraction stages is unusual because of the large solvent volumes and expensive equipment required. Special centrifuges which contain mixing and settling sections may be used to carry out several stages simultaneously. Continuous extraction methods, either with columns, steady state countercurrent centrifuges, or mixer-settler separators are frequently used to decrease solvent requirements and increase yields. Group separations cannot give a 99% yield of pure substance in one or two extraction stages unless all impurities have β values of 100 or greater and the substance has a high partition ratio for the extracting solvent. The case for two substances with a β value of 9 has been discussed previously.

A. Solvent Properties

In investigating solvents suited for a particular compound or isolation step the following criteria should be considered.

1. Partition Ratio

The substance should favor the *added* solvent. If a single phase is extracted with successive portions of solvent the fraction remaining after n

extractions is q^n in which q is the fraction not removed by a single extraction, $1/(K + 1)$. Thus, if a solvent having a partition ratio of 1 is used in equal volume for four successive extractions the fraction not extracted is 6.25% ($0.5^4 \times 100$). More efficient utlization of the same amount of solvent (less efficient utilization of time) is obtained if more extractions are utilized. By using eight half volumes the fraction not extracted is about 4% ($0.67^8 \times 100$).

These calculations emphasize the desirability of having a high partition ratio for group separation processes. Low ratios require excessive amounts of solvent and a large number of extractions for high yields. In addition, it is likely that selectivity will be poor under conditions of low partition ratio. On the other hand, high partition ratios (100 can frequently be realized) give high yields with little solvent and effort and may have good selectivity.

2. *Solvent Polarity*

The extracting solvent should be as unlike the extracted solvent as is consistent with a good partition ratio. In general, if the solvent used has nearly the same polarity as the extracted solvent, poor selectivity is obtained in comparison with a solvent having less similarity. This may not be true in cases in which the solvent has some special complexing, hydrogen bonding, or sequestering property which is unique for the substance being extracted.

3. *Solvent Removal and Loss*

The solvent should be readily removed by one of the methods outlined below. Loss of the solvent by entrainment, by excessive solubility in the extracted phase, or by volatility should be avoided if possible.

4. *Reactivity*

Reactive solvents should be avoided if other solvents having similar partition ratios can be used. Ester solvents are hydrolyzed by strongly acidic or basic solutions and can react with both alcohols and acids by transesterification. Free amine groups can be acylated under certain conditions with esters or can form Schiff's bases with ketonic solvents. Ethers should not be exposed to conditions conducive to peroxide formation and should be avoided if traces of peroxide would damage the product. This is especially true for peptides containing sulfur amino acids. Strongly alkaline conditions can cause the decomposition of aliphatic halogen compounds. Alcohols should not be used under conditions which could lead to esterification of carboxyl-containing products.

5. *Ease of Separation*

Emulsion formation and lack of ready separation of solvents may result if the density of the two phases is quite similar. Some solutes encourage emulsification and the prevention of such emulsions may require exacting conditions of temperature, salt concentration, and surfactant addition. Certain solvents can be mixed in liquid form and then frozen for separation with little loss in efficiency. Halogenated phenols and aromatic hydrocarbons have been used in the solid state (*13*). The solid extractant can then be separated by filtration. Solid solvents which are incapable of dissolving in the extracted phase, e.g., cross-linked polymers, have been used in the large-scale recovery of natural products. Polymers which have been used are carboxylic acid ion exchange resins and specially formulated gels, such as Duolite S-30, Permutit DR, XAD-2, and Sephadex gels. Recovery from these polymers usually involves a change in the composition of either the resin or liquid phase.

B. Solvent Removal Methods

Once the material has been transferred to the selected solvent it must be recovered by some economical means. There are several methods for accomplishing this.

1. Evaporation

The solvent extract can be evaporated to dryness or concentrated to a volume such that the product can be recovered by precipitation or extraction.

2. Back Extraction

The solvent can be removed by extracting the substance into another solvent. Back extraction is usually accompanied by additional purification. Reextraction into the orginal solvent without considerable dilution requires a change in the partition ratio. This can be accomplished by adding a second solvent, or, if the desired substance is an acid or base, by changing the pH. The effect of a second solvent on the partition ratio has been mentioned.

For ionizable substances it can be assumed that there will be a large difference in the partition ratio of the nonionized form in comparison with the ionized form. This is especially true if the solvents being used have a large difference in polarity and if the counterion of the ionized form is highly insoluble in the nonpolar phase. If one of the phases is water the fraction ionized in the aqueous phase can be approximated from the Henderson-Hasselbach equation [Eq. (II.1)].

$$\frac{\log \text{ionized}(\alpha)}{\log \text{nonionized}\ (1-\alpha)} = \text{pH} - \text{p}K$$

Thus, at two pH units from the pK the substance is either 99% ionized or 99% nonionized. If the steps of extraction and back extraction are operated at a pH spread of three units nearly maximum efficiency can be obtained.

Ordinarily the pK is within the operating limits. However, the inherent partition ratio of the nonionized form for any particular substance determines the best pH range.

Stability considerations may indicate a necessity for operation within a certain pH range and influence the solvent chosen. The effective partition ratio as a function of the inherent partition of the nonionized form and pH relative to pK for singly ionized substances is plotted in Fig. II.3.

Benzylpenicillin is an organic acid with a pK of 2.7. At 0°C and pH 2, the approximate half-life is 3 hr; at pH 3, 15 hr; and at pH 4, 180 hr (*14*).

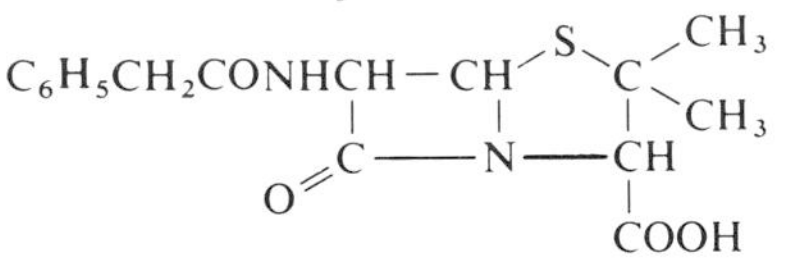

$$C_6H_5CH_2CONHCH-CH \langle S \rangle C(CH_3)_2 ; \; O{=}C—N—CH(COOH)$$

Benzylpenicillin

In a comparison of two solvents, amyl acetate and methylcyclohexanone, which have inherent partition ratios of 21 and 185, respectively, for the free acid, it was observed that the ionized alkali metal salts have very low partition ratios. Under these circumstances, the partition ratio calculated from Eq. (II.2) agrees with the value obtained experimentally over the pH range 2.5–7.5. Typical calculated values and the observed values for benzylpenicillin are plotted in Fig. II.5. A 95% extraction yield into an equal volume of solvent requires a partition ratio of 19. This is obtained at pH 2 with amyl acetate and at pH 3.7 with methylcyclohexanone. At pH 6.0 and at pH 6.7 the yields on back extraction into an equal volume of water are 99% and 98%, respectively. Thus, at equal efficiency the solvent with the higher inherent partition ratio can be operated at 1.7 pH units higher. Using this solvent the possible overall recovery of penicillin is higher because losses due to instability are less regardless of the time and temperature of the extraction.

In the above case the partition ratio of the ionized form is very low. This may not be true when polar solvents containing appreciable quantities of water are used. In addition, salt forms, perhaps only partly ionic, or inorganic counterions and organic molecules having a large nonpolar moiety, or salt forms of organic acids and bases may have appreciable partition ratios into the less polar phase. These conditions will detract from the efficiency of back extraction processes which depend on a pH change. The effects of ionizable substances which have a high partition ratio for the less polar phase have been mentioned.

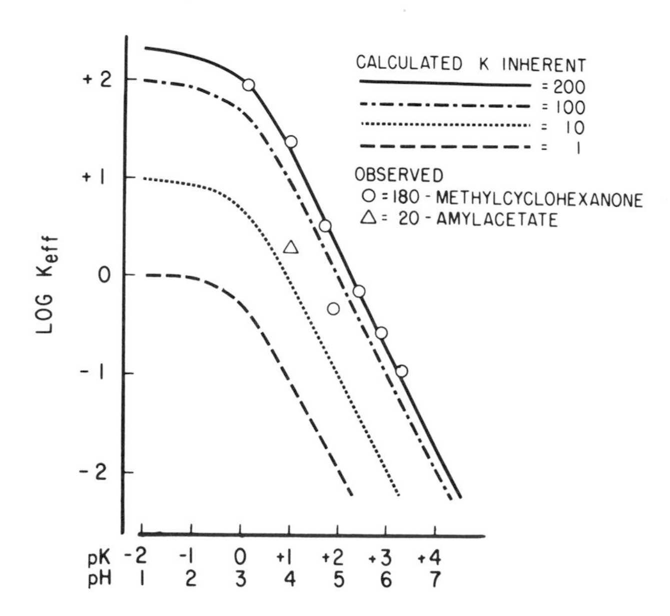

FIG. III.5. Calculated and observed partition ratio of benzylpenicillin.

Streptomycin, a strongly basic antibiotic containing two guanido groups, can be extracted using polar solvents, such as *n*-butanol, containing solvent-seeking acids such as lauryl sulfate, stearic acid, and the like. Similarly solvent-seeking amines can be used as carriers for strong acids.

3. *Adsorption*

The solvent can be removed by adsorption of the solute on a suitable adsorbent. This is especially desirable if additional purification is needed and a good adsorbent is known for the desired substance. Vitamin B_{12}, a neutral compound, is extracted by benzyl alcohol, a high-boiling solvent. Although the vitamin can be recovered by adding a nonpolar solvent and extracting into water, this gives little purification. Highly purified product can be recovered by adsorbing the vitamin on activated alumina and eluting with methanol. Similarly, Florisil has been used for the recovery of the tetracycline antibiotics from solvent extracts.

Ionizable substances can be removed from solvents by ion exchange methods. As a general rule solvent-extractable organic substances are difficult to recover from ion exchange resins. For this reason it is usually preferable to use ion exchange alone if high recovery can be obtained.

4. *Precipitation*

The addition of a solvent in which the product is insoluble can be used in rare cases. Usually it is not possible to use a precipitation method directly on a solvent extract because of the large volume of solvent required and the probability that a low yield will be obtained due to solubility of the extracted material. Some selectivity and high yield may be obtained by conversion to an insoluble salt and precipitation or by conversion to a salt of lower solubility by metathesis. Since many amorphous salts are hygroscopic, it is usually desirable to dry the solvent solution before adding the precipitating agent.

If the reagent used for the precipitation yields a crystalline product, this procedure can be quite useful in obtaining additional fractionation. A study of reagents which might yield crystalline salts should always be considered. Although it is difficult to predict the specific reagent which is most useful for a particular compound, little fractionation and decreased likelihood of crystallization can be expected if the reagent produces a drastic change in solubility. Organic reagents frequently form salts which have desirable solvent solubility properties. In addition an excess of reagent can be added without the danger of producing a drastic pH change which would occur with even slight excesses of mineral acid or fixed alkali. Slight excesses of these reagents can be quite destructive in solutions which are nearly anhydrous.

Benzylpenicillin is recovered in high yield from ester solvents by neutralization with various amines, such as triethylamine or *N*-ethylpiperidine. The tetracyclines may be precipitated from butanol as quaternary amine derivatives.

III. APPLICATION OF SOLVENT EXTRACTION FOR FRACTIONATION SEPARATION

Substances having low β values can be separated by solvent extraction procedures using countercurrent distribution (CCD), steady state extraction methods, or partition chromatography. Each method is frequently applicable to the same separation problem but each has advantages which may indicate that one is preferable for the separation being studied. Some of the factors to be considered and a partial comparison of the methods are presented in Table III.2.

IV. PARTITION CHROMATOGRAPHY

Partition chromatography is a fractionation procedure utilizing an immobilized liquid phase in a column. A mobile solvent, immiscible with the immobile phase, is passed through the column. This procedure differs from a group separation process employing continuous extraction of a stationary liquid phase in a column in the distribution of the solute in the immobilized liquid. In this case, the solute is distributed through the entire immobile

TABLE III.2

COMPARISON OF PARTITION CHROMATOGRAPHY, COUNTERCURRENT DISTRIBUTION, AND STEADY STATE EXTRACTION

	Partition chromatography	CCD	Steady state extraction
β value, minimum practical	1.15	1.2	1.2
Number of stages, practical	1000–6000	500–2000	100
Amount charged	Low because stationary volume limited	Better but still low	Large
Operating time	1–3 days	2–7 days	Continuous
Purity indication	Possible	Precise	Indecisive
Dilution of product	Very high dilution	Relatively low dilution sometimes possible	1/10–1/100
Equipment availability			
Cost	Good, inexpensive	Poor, expensive	Poor, expensive
Automation	Good	Good	Good
Continuous monitory	Possible	Difficult	Difficult
Space required	Small	Major	Major
Mechanical problems	Primarily temperature and flow rate regulation	Complex equipment	Complex
Solvent limitations	Wide variety possible	Both partition ratio and emulsification must be controlled	Same as CCD
Yield	Good	Good	Good

phase at the beginning of the process. On the other hand partition chromatography can be considered as a series of individual columns of stationary phase, each having a unique composition.

The classic studies of Martin and Synge (*15*) provided the first examples of such columns and pointed out the relationship of the fractionation process with a series of sequential columns. The term, liquid–liquid chromatography (LLC), is perhaps more descriptive of the process and is sometimes used.

A solid support is used which fixes the immobile phase in position. This solid does not participate in the process and ideally should not exert any influence on the partition ratio. Precipitated silica was used as the solid support in the orginal work and the polar phase was stationary. Partition chromatography differs from chromatography with solid adsorbents in that the position of a band of solute is determined by its partition ratio between the two liquid phases and the relative volume of the phases employed.

The successful operation of a partition chromatography process depends on the selection of a suitable support, a suitable solvent system, and careful control of the mechanical features of the process.

A. Properties of the Solid Support

An ideal support would have the following properties:

Physical

(1) The individual particles should have uniform size and shape.

(2) The shape should be such that pools of "stagnant" or inaccessible liquid do not form or exist in either the mobile or nonmobile phases.

(3) Mobile and immobile phases should be in intimate contact in "thin" layers.

(4) The support must be stable, i.e., not change form or size, during the chromatographic process.

(5) A substantial quantity of the immobile phase should be retained.

Chemical

(1) Adsorption of the solute should not take place.

(2) The substance should be both nonreactive and noncatalytic.

(3) The surface should retain only the immobile phase.

(4) Strong forces of solvation, hydrogen bonding, etc., which would result in a change of the solvent properties of the immobile phase depending on its physical location relative to the surface should not exist.

Actually there is no completely ideal support. Commonly used supports are starch, cellulose fiber, diatomaceous earth, silica, and pulverized fire brick for polar stationary phases and rubber and silanized diatomaceous earth or cellulose for hydrophobic stationary phases. Recently certain organic gels of the cross-linked dextran type have been used as supports for both polar and nonpolar phases. Cellulose fiber and diatomaceous earth have been found satisfactory for immobilizing capryl alcohol and similar solvents in the presence of a more polar mobile phase.

B. Properties of the Solvent System

Partition chromatography requires the availability of a suitable solvent system. A number of properties of the solvents used contribute to the success or failure of the approach. Some of the most important of these are discussed below.

1. Stability

The components of the system should be chemically unreactive with each other, with the support, and with the solutes being partitioned. In general use of buffered systems containing esters should be avoided as slow hydrolysis of the ester eventually affects the pH of the system.

Although good operating procedure is to maintain constant temperature, this is not always possible. Some solvent systems are more susceptible to changes of composition with temperature and should be avoided if possible.

Phase ratios should be relatively insensitive to solute concentration. Some systems, especially those containing water in both the polar and nonpolar phase, shift composition markedly in the presence of water-seeking solutes.

Buffer systems should have sufficient buffering capacity to maintain constant pH in the presence of the highest concentration of solute which will be present during the run.

2. Partition Ratio

The partition ratio should be in favor of the immobile phase. Usually the volume of the immobile phase is lower than that of the mobile phase by a factor of 2–5. In the absence of other influences, the separation increases the longer the desired substance remains in the column, i.e., the greater the partition ratio is in favor of the immobile phase. This effect is counterbalanced by increased time, larger solvent volume, and diffusional spreading of the zone. Even in the absence of diffusion the increase in separation ability achieved with large elution volumes becomes unattractive beyond an effective distribution ratio, G, of about 0.1. In this case, the effective distribution ratio, G, is equal to the ratio of solute in the mobile phase to that in the stationary phase in any column cross section.

If it is assumed that all substances have the same plate height and that this does not change with retention volume, substitution of G and βG for p_a an p_b and rearranging expression (I.18) gives the following relationship:

$$r^{\frac{1}{2}} = R_r \frac{\beta(2G + 1) + 1}{\beta - 1} \qquad \text{(III.7)}$$

in which r is the total number of plates in the chromatographic column. Typical values obtained at $R_r = 2$ are illustrated in Fig. III.6.

The practical operating ranges of partition ratio and column height are indicated by the dashed lines. This figure reveals the effect of increasing column efficiency and changing selectivity, β, on the separation process. It is apparent that for any value of β there is little to be gained in decreasing the effective distribution coefficient below about 0.1. Since a decrease from 0.1 to 0.01 requires an increase of almost 10 times in the solvent and time required for a single column it is apparent that a study of the factors influencing the selectivity would probably be more productive that carrying out a separation using a poor distribution ratio. An average well-operated column of 100-cm length would have 400–600 theoretical plates and usually would be operated

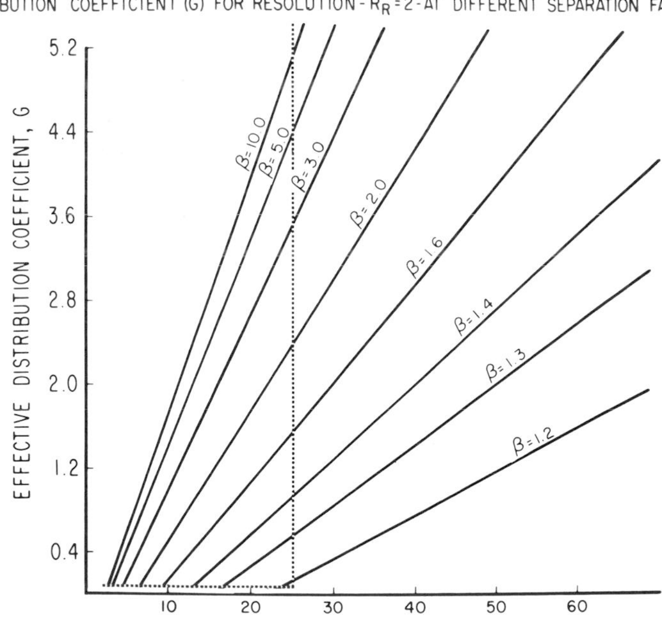

FIG. III.6. Resolution in columns. Relationship of distribution coefficient, G, separation factor, β, and the number of effective plates, r.

a at G of 0.1–2.0. The type of separation possible is seen in Fig. III.6. The maximum efficiency is in the range of β values of 1.2–1.4.

3. Factors Influencing Column Load

A significant proportion of the total solvent in any cross section of the column should be immobile phase. This permits the use of greater amounts of solute per operating volume. In addition the solvent system should readily dissolve the solute, thus permitting higher concentrations of solute to be used without affecting the partition ratio. It has been stated (*16*) that the highest possible ratio of immobile phase should be employed to eliminate adsorption effects of the solid support.

C. Mechanical Problems

After the system has been selected it must be operated mechanically. Chromatography is an inherently simple process requiring only a column and a method for collecting fractions and analyzing the effluent stream from the column. However the efficiency of the operation depends on careful attention to details of packing, charging, development, and analysis. Partition chromatography is especially sensitive to poor packing, lack of equilibration and poor application of the charge. The reasons for this are brought out in the requirements of a perfect support and solvent system mentioned above.

Frequently both paper and thin-layer chromatography are mechanistically partition chromatography. The utilization of a relatively thin section in vapor saturated with solvent provides standardized conditions. Lack of equilibration with the developing solvent may not prevent separation in these cases.

D. Mechanism of Separation

If a true partition or liquid–liquid chromatographic mechanism is operating, it is apparent from the discussion in Chapter I that the degree of separation obtained in a particular system is dependent only on the number of stages or theoretical plates in the column. While the number of theoretical plates obtained with a particular system and column may be different with different compounds due to different diffusion coefficients and rates of transfer from phase to phase, the maximum number of plates can be influenced markedly by the physical and chemical properties of the system. The "ideal" circumstances detailed above are those which lead to maximum performance.

From Eq. (I.16) it is seen that the band spread is proportional to $1/r^{\frac{1}{2}}$ in which r is the number of theoretical stages in the column. Since the retention volume, V_e, is not influenced by r but only by the distribution coefficient, G, the importance of maximizing r is obvious.

1. Factors Influencing Plate Height in Columns

Recent studies by Giddings (*17*), Klinkenberg and Sjenitzer (*18*) and Van Deemter *et al.* (*19*) have emphasized many of the important features which influence the height of a theoretical plate (HETP) in gas–liquid chromatographic columns. A number of these factors are also important in partition chromatography.

A. LONGITUDINAL DIFFUSION. In the ideal chromatographic separation all of the molecules of a particular substance would have the same rate of migration through the column. However, the process of diffusion clearly prevents this from happening since at any time during a chromatographic run the molecules of solute are diffusing in all directions from the migrating zone. Some molecules will therefore travel faster and others slower than the main band.

A column containing only liquid could be set up. If a thin layer of the same liquid containing a solute were placed at the top of the column and then fresh liquid added at the top while withdrawing liquid from the bottom in such a way that no distortion of the layer of solute resulted from density differences, wall effects, temperature convections and the like, the solute zone would produce a concentration profile within the column having the shape of a Gaussian curve, if the time required for displacing the liquid from the column is less than the time required for the solute to diffuse equally into all parts of the liquid. In this ideal situation the band spread would be determined only by the time of displacement and the rate of diffusion of the solute. A very rapid displacement would produce a sharp zone and slow development, a broad zone.

Solutes having different rates of diffusion produce different band spreads for the same retention time. Diffusion coefficients have the dimensions (length^2) (time^{-1}), and are influenced by the shape and mass of the substance as well as the viscosity of the medium. In water, the diffusion coefficient of KCl is about 1.5×10^{-5} cm^2/sec and that of sucrose is about 0.4×10^{-5} cm^2/sec at room temperature.

The variance, σ, due to diffusion has been given by Giddings as

$$\sigma = (2Dt)^{\frac{1}{2}} \qquad \text{(III.8)}$$

in which D is the diffusion coefficient of the solute in the medium employed and t is the time of development. If two bands of solute were applied to the column, the linear distance the solvent front must travel before the second band could be applied and remain separated from the first can be calculated from the above equation. In 1 hr, σ for KCl is 3.3 mm and the 4σ band spread is then ± 6.6 mm, from the original band center. If 24 hr is required to develop the column, 95% of the KCl, applied originally as a 1-mm band, would emerge in a band 64.5 mm wide. A second zone could not be separated unless applied after the original zone had traveled 64.5 mm.

Diffusion coefficients diminish as molecular size increases (in 1 hour, σ for sucrose is 1.7 mm). For 24-hr development a second zone of sucrose could be applied after the first zone had traveled only 16.6 mm.

In ordinary chromatography the above values represent a maximum effect since at any time a considerable portion of the moving zone is confined to the stationary phase which has a different, usually much lower, diffusion coefficient.

B. EDDY AND LATERAL DIFFUSION. With the flow rates normally used in chromatography the liquid moves through the column under conditions of viscous flow and the velocity of the liquid film at the walls of the column is less than the velocity in the main stream. This factor causes an additional band spread since the moving band encounters liquid near the wall of the column which has a lower solute content than the main band. This is compensated to some extent by lateral diffusion into and out of this zone, which tends to deplete the concentration at the front of the zone and increase it at the rear. Thus lateral diffusion tends to reduce the band spread due to differential cross-sectional velocity.

If the column is packed with an inert solid, this solid provides new surfaces which may retard the flow of liquid similarly to the wall of the column. It is also apparent that individual segments of liquid will move through the column at different linear velocities since the solid will cause areas of constriction and expansion. In addition some parts of the liquid will be forced to take a longer path around obstructing particles than other parts of the liquid. Quite obviously these factors will be influenced by the uniformity of size and shape of the solid particles. If an individual molecule is "locked" into a particular segment of liquid it would move through the column at the same velocity as its liquid carrier. Some molecules would thus move faster than the average and some slower, with the net effect being increased band spread. The term "eddy diffusion" has been used to describe this property in gas–liquid chromatography. For liquid chromatography under the usual flow conditions this factor is independent of the actual velocity of the liquid through the column.

Since, in fact, solute molecules diffuse from one portion of liquid to another and would have a net movement toward liquid which is relatively depleted, lateral diffusion (i.e., diffusion not in the direction of net flow) decreases the band spread due to eddy diffusion. At low flow rates, band spread is increased by diffusion along the axis of the column whereas lateral diffusion minimizes the overall effect of eddy diffusion. As the velocity increases the ability of lateral diffusion to overcome the spreading effect of eddy diffusion becomes less. The net variance for the combined eddy and lateral diffusion as described by Giddings is as follows:

$$\sigma = \left(\tfrac{1}{2}\lambda\, d_p + \frac{D}{Wv\, d_p^{\,2}}\right)^{-\frac{1}{2}} \tag{III.9}$$

in which d_p is the diameter of the solid particle, v is the velocity of flow, and λ and W are factors due to the shape, uniformity, and packing of the particles. This equation predicts an optimum flow rate for a particular column which depends on the interaction of the solid particles, flow velocity, and diffusion coefficient of the solute. Different maxima are obtained for solutes having different diffusion coefficients.

c. Internal Diffusion. The discussion above has not considered the effects of events within the solid particles. In partition chromatography these particles contain the immobile-phase liquid. In considering the effect of this factor on the band spread, the importance of diffusion of the solute across the phase boundary and within the stationary phase becomes apparent. If the immobile phase is present in perfectly uniform shape and size with constant surface-to-volume ratios, the band spread is increased by some factor due to the barrier of the liquid film to free diffusion of the solute. The barrier can be visualized as due to unique properties of the liquid film and probable orientation of solvent molecules at the interface. If one or both of the phases contain buffers the effective pH of the interface may be different than the bulk liquid. This can act as an additional barrier to diffusion, i.e., diffusion is facilitated in one direction. For some solvent systems and solutes the barrier effect can be substantial. The net result is to increase the band spread by some factor related to the total area of the film. If the effect is substantial, the requirement for uniform shape and volume of immobile liquid particles becomes greater.

A second band spreading effect due to the immobile phase is the actual diffusion itself. The solute must diffuse in and out of the immobile phase at each stage of the column. Since some solute molecules will diffuse further away from the interface than others, this will cause an increase in band spread which is related to the thickness of the immobile liquid and the diffusion coefficient of the solute in the immobile phase. If the composition of the immobile phase is not uniform due to interaction with the surface of the support medium then the distribution of solute molecules in the immobile phase may likewise be nonuniform.

Chromatography theory cannot reasonably predict the effects of all these factors on band spread. However, the net band spread, which is a function of the system and column packing, etc., can be measured and resulting values can be used to determine the application of a particular system. The procedures for doing this have already been discussed.

D. NONLINEAR DISTRIBUTION COEFFICIENTS. In all of the above discussion it has been assumed that the partition ratio of the solute had remained constant during the migration of the band of solute from the top to the bottom of the column. Since the concentration of the solute decreases as the band spread increases and since separation of the solutes in a band originally applied to the column is taking place during the chromatography, the partition ratio may change during the separation. These changes do not influence the number of theoretical stages in chromatography. However, the changing distribution coefficient affects band spread and causes skewing. If the distribution coefficient shifts in favor of the mobile phase as the concentration diminishes, the net result is a diffuse leading edge of the band since the more dilute regions in front of and behind the main zone speed up. If the shift is in favor of the immobile phase the front becomes sharper and the rear more diffuse. Since the partition ratio change is apt to be less at low concentration, this may limit the amount of material which can be charged to a column.

E. Applications of Partition Chromatography

The method has been applied to a variety of substances of greatly different polarity. Since the original applications of Martin were with compounds of medium polarity, a polar stationary phase was used. Later, in applications to nonpolar substances, a nonpolar stationary phase was employed. This type of partition has been termed "reverse" phase partition chromatography.

Many applications of partition chromatography were designed for analytical purposes using milligrams of material. This use has been largely superseded by gas–liquid chromatography, which is generally much faster, more precise, and more sensitive. If the desired substance is not volatile or cannot be converted to a volatile derivative or if the recovery of sufficient material for chemical characterization or biological testing is the primary objective, partition chromatography may remain the method of choice. For the preparation of gram quantities of pure substances very large columns and large quantities of solvents are required.

As a general rule, if a purification factor of greater than 50-fold is required some additional enrichment is desirable before attempting a partition chromatogram.

The limits of obtainable separation are indicated in Fig. III.6. This has been calculated for the separation of two compounds allowing 2.5% of each component in the overlap region. If two substances are present in unequal quantities and if the desired component is the minor substance, it is apparent that a much larger degree of separation may be necessary.

1. Insulin

The recent work of Slobin and Carpenter (*20*) in the separation of insulin (a peptide containing 52 amino acids) and monodesamidoinsulins, the compounds resulting from the conversion of $-CONH_2$ to $-COOH$, is indicative of the resolving power obtained with well-operated partition columns. Samples of crystalline zinc insulin were shown to contain up to four compounds. Since insulin contains four asparagine residues, these may be different monodesamido derivatives. The same compounds were produced by allowing insulin to stand in 0.1 *N* hydrochloric acid at 40°C. In one system the partition ratios observed were as follows: insulin, 0.104; component II, 0.122; component III, 0.134 and component IV, 0.160. The separation factor, β, relative to insulin for the three compounds is then, for component II, 1.17; component III, 1.29; and component IV, 1.53. The systems used were not capable of separating all factors simultaneously. However, good separation of insulin and component III was usually obtained. Columns suitable for the chromatography of 200 mg of material are about 3.75×40 cm. The time required for development is about 16–24 hr. The apparent number of theoretical plates is about 10–12 per centimeter. By selection of the type of support it was found possible to obtain a ratio of mobile to stationary phase of about 2. Since the partition ratio of insulin in the system employed is about 0.1, the separation was operated at a distribution ratio of about 0.2. Referring to Fig. III.6, it can be seen that with 400 plates per column ($r^{\frac{1}{2}}$ = 20) and a G of 0.2, a relative resolution, R_r, of 2 for compounds with a β of 1.2 cannot be obtained. It was, in fact, observed that the four compounds could not be prepared pure in any single run. These studies also indicate the importance of stability information when obtainable. In this case, with the 0.1 N hydrochloric acid system it was impossible to prepare pure insulin, as the acid hydrolysis products were always present in even small fractions of the peak material. In spite of this instability, these methods remain the best for pure insulin preparation since no ion exchange method has been found which gives similar yields. Countercurrent extraction requires more time and produces even greater contamination with hydrolysis products.

2. *Aldosterone*

It is sometimes possible to achieve a greater separation of many components from a single chromatographic run by using a gradient elution procedure. This was used in the isolation of aldosterone (*21*). The structural formulas of the related steroids indicate the problem encountered.

Since all three steroids have the same carbon-to-oxygen ratio and three of the oxygens are identical, the differences in polarity are due to the ability of the remaining oxygen atoms to form hydrogen bonding complexes. The ability of HO groups to enter into hydrogen bond complexes either as donors

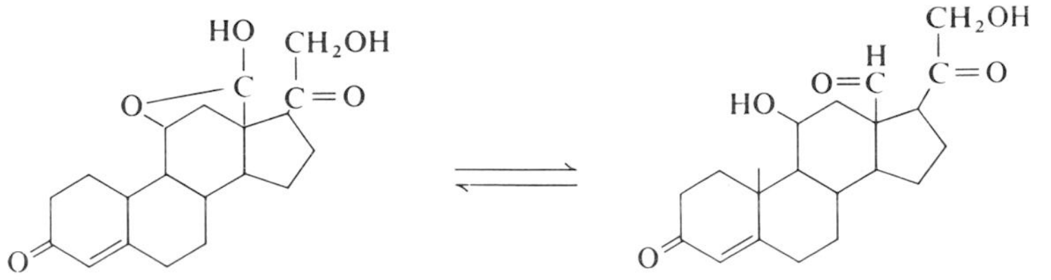

Aldosterone

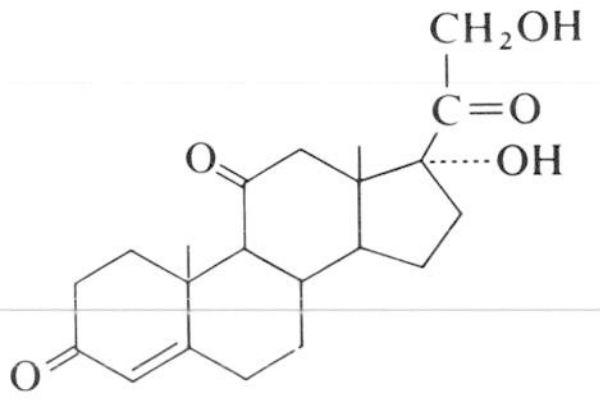

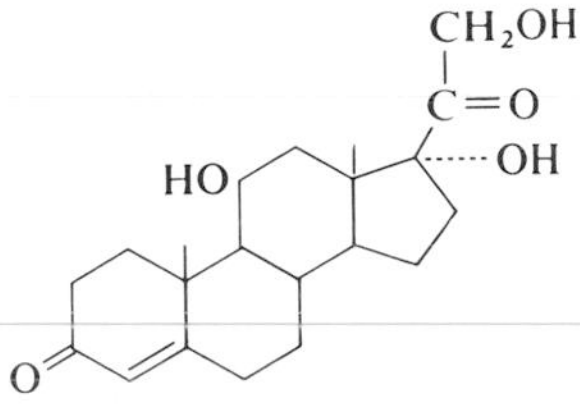

Cortisone

17-Hydroxycorticosterone

or acceptors, whereas carbonyl or ether oxygens can only act as acceptors, causes hydrocortisone to be the most polar of the three compounds. Similar reasoning leads to the expectation that the acetal form of aldosterone would be the least polar. Using paper chromatography with a system of propylene-glycol/toluene, aldosterone does, indeed, have the greatest R_f but is only slightly separated from cortisone. Hydrocortisone is the most polar. However, with another system, methanol, water–toluene, ethyl acetate, aldosterone is less mobile than cortisone and is only poorly separated from hydrocortisone. This unexpected reversal in mobility provides a method of separating and identifying aldosterone if the two systems are used in sequence. This method was used to detect aldosterone in column fractions.

For the isolation of aldosterone (*21*), 500 kg of beef adrenals was extracted with solvent to yield 167 g of solvent-soluble material. A group separation procedure with petroleum ether and aqueous methanol eliminated nonpolar contaminants and yielded 27 g of concentrate for partition chromatography.

A column was prepared using 1.4 liter of water and 1.4 kg of purified kieselguhr and packed with petroleum ether as the mobile phase. The crude concentrate was mixed with kieselguhr, dried, and applied as a solid admix to the top of the column. Fractions of about 1.5 liters were collected at the rate of about two each day. The eluant was gradually increased in polarity by the addition of benzene. Each fraction was analyzed by paper chromatography and the total mass determined. Since the increase in polarity of compounds coming off the column was readily seen, the fractions containing aldosterone were easily located. This occurred at a benzene content of about 75%. Fractions 86–90 contained most of the aldosterone. The total weight, 357 mg, was estimated from paper chromatographic analysis to contain about 10% aldosterone. Thus, the degree of purification was about 75-fold. Cortisone was obtained from fractions just prior to aldosterone and hydrocortisone appeared just after the hormone.

The final chromatography was carried out using a partition column of cellulose. The immobile phase was prepared from a mixture of toluene, petroleum ether, methanol, and water (66:33:60:40). A column containing 60 g of cellulose and 30 ml of lower phase was used to chromatograph 324 mg of the fraction obtained above. A solvent change to a 3 to 1 toluene–petroleum ether mixture was applied near the end of the chromatography. The aldosterone-containing fractions (about 200 ml) yielded 22.5 mg of crystalline compound from the 48 mg (sevenfold purification) of total mass.

Once an unknown substance has been obtained for the first time and carefully characterized, the isolation procedure can usually be improved. This is illustrated by the use of an acetylation procedure for aldosterone isolation. The mixture of hydrocortisone and aldosterone obtained from the partition column yields a monoacetyl derivative of aldosterone and a diacetyl derivative of hydrocortisone. These compounds are readily separated and the aldosterone regenerated.

3. *Use of Hydrophilic Gels*

It is potentially advantageous to have a large volume of stationary phase since this results in improved capacity of the partition system. Spherical particles occupy about 60–64% of the total volume. Synthetic gel matrixes of the type used for gel-permeation chromatography may allow a 1:1 volume ratio for stationary to mobile phase. Both polar and nonpolar gels are available. With hydrophilic gels the change in volume which occurs as a function of hydration and temperature requires careful control. A system containing a dextran gel was used for the fractionation of actinomycins.

A. ACTINOMYCIN. The actinomycins are antibiotics containing cyclic peptide groups. A number of different compounds are usually produced simultaneously. The compounds cocrystallize and paper chromatography of crystalline products frequently indicates a mixture. Since the pure compounds have different antibiotic potency and toxicity, therapeutic use of the substances requires high purity. Actinomycin C was shown to be a mixture of three compounds, C_1, C_2, and C_3, in order of increasing polarity. These compounds have the structures indicated:

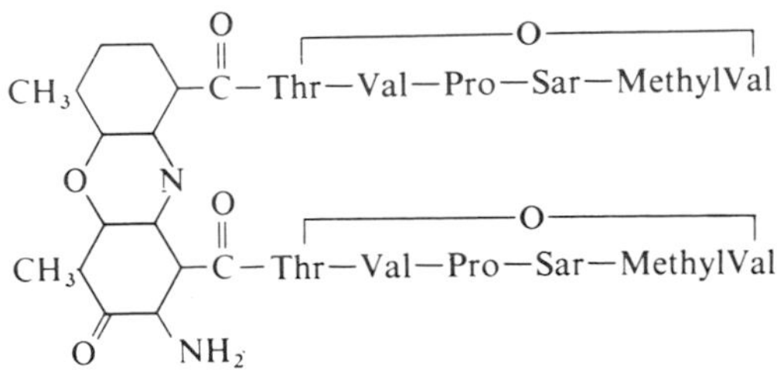

O
C—Thr—Val—Pro—Sar—MethylVal
CH_3
O N
O
C—Thr—Val—Pro—Sar—MethylVal
CH_3
O NH_2

Actinomycin C_1

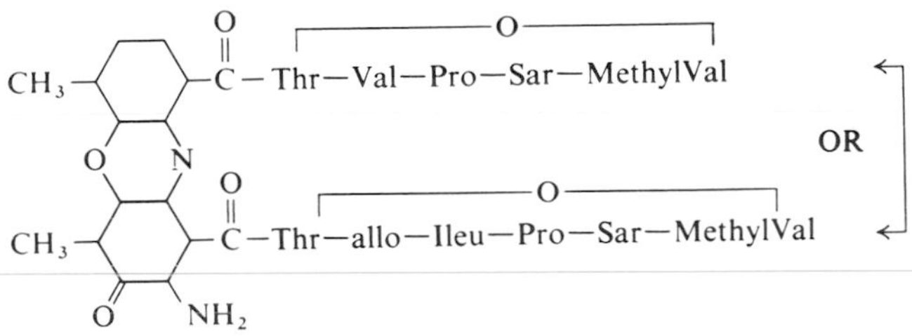

O
C—Thr—Val—Pro—Sar—MethylVal
CH_3
O N OR
O
C—Thr—allo—Ileu—Pro—Sar—MethylVal
CH_3
O NH_2

Actinomycin C_2

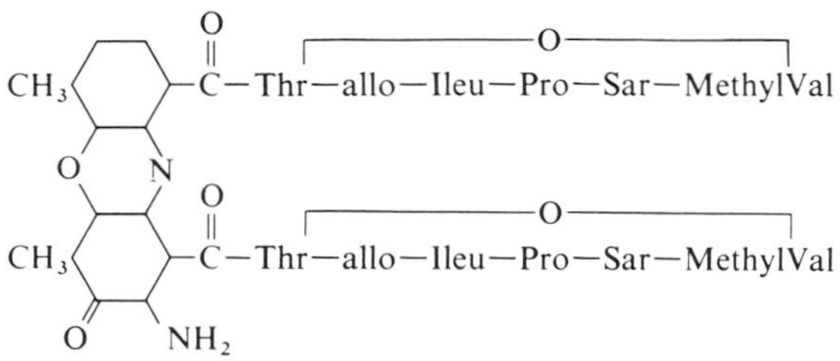

```
                 O          ┌─────────────O─────────────┐
                 ||         
CH3-(ring)-C—Thr—allo—Ileu—Pro—Sar—MethylVal
        O    N
                 O          ┌─────────────O─────────────┐
                 ||
CH3-(ring)-C—Thr—allo—Ileu—Pro—Sar—MethylVal
        O    NH2
```

Actinomycin C_3

The compounds are separated in a system prepared from di-n-butyl ether, *n*-butanol, and 9% aqueous sodium *m*-cresotinate. The actual partition ratio is influenced by the ratio of solvents, i.e., increasing the proportion of butanol increases the partition ratio into the solvent phase. In the absence of the salt the partition is strongly in favor of the solvent phase. Variants of this system have been used with either Celite or cellulose support to prepare small quantities of pure actinomycins. Schmidt-Kastner (*22*) reported the use of Sephadex G-25 for the partition column support. With a 3:2:5 volume ratio of the above solvents, 320 mg of a mixture was clearly separated on a 40-cm column containing about 380 ml of bed. The approximate distribution coefficients were 1.2, 0.3, and 0.1, respectively.

B. CARBOHYDRATES. An interesting use of cross-linked dextran in which the mechanism of separation appears to be partition between liquid phases of different composition has been described by Zeleznick (*23*). In this case the dry gel was treated with a ternary mixture: *n*-butanol, acetic acid, and water (65:15:25). This mixture is very near the composition of the upper phase obtained from the two-phases 4:1:5 system. Hence the addition of a small amount of water causes the separation of two phases. Columns prepared with this solvent mixture and Sephadex G-25 gave a retention volume of glucose of six to nine column volumes. The stationary phase appears more polar than the mobile phase since rhamnose, *N*-acetylglucosamine, glucose, and glucosamine hydrochloride are eluted in order.

4. Miscellaneous Applications

Other examples of partition chromatography are given in Table III.3. These have been selected to illustrate the general areas of usefulness of the method and to show conditions which give good separation. In addition the column operating parameters are noted when available. These include flow rate, amount of material charged, effective plate height, and efficiency.

The examples given in Table III.3 are sufficient to indicate the variety of systems which can be employed in partition chromatography and its application to different types of compounds.

The procedure has also been used for the following separations: C_6–C_{12} fatty acids (*39*); C_{12}–C_{18} fatty acids (*40*); C_6–C_{20} fatty acids (*41*); C_8–C_{20} fatty acids (*24*); Fatty acids (*42, 43, 44, 45, 46*); lipids (*47*); iodophenylsulfonylamino acids (*48*); dinitrophenyl amino acids (*49*); peptides, insulin (*50, 51*); hypertensin (*53*); proteins, globulin (*52*); glycosides (*54*); methylated sugars (*55*); carbohydrates (*22*); polysaccharides (*56*); steroids (*57, 27*); bile acids (*58*); insecticides (*59*).

V. COUNTERCURRENT DISTRIBUTION

The general procedure for countercurrent distribution (CCD) has been discussed and a brief comparison with partition chromatography is given in Table III.2. Separation by CCD is a valuable procedure for the preparation of sufficient quantities (1–50 g) of pure material for chemical characterization. The pioneering work of Craig in the design of equipment enabled the wide-scale application in the laboratory. The method is precise and the results obtained are readily interpreted. For analytical use computation of band spread and fitting of experimental data to a theoretical curve can detect impurities in quantities as little as 0.01% or the presence of a major impurity having a β of as little as 1.00035.

A. Physical and Chemical Requirements

1. Solvent System

A satisfactory CCD separation requires an immiscible solvent system which has the following properties:

(a) Readily dissolves the substance in both phases
(b) Does not tend to emulsify and separates rapidly
(c) Has a constant partition ratio at the concentration which will be used
(d) Is easily removed
(e) Has a partition ratio for the desired component of about 0.2–5
(f) Permits a good separation factor
(g) Does not react either with itself or any of the materials to be separated

2. Chemical Stability

The desired component must be stable under the conditions of the distribution. In addition it is desirable if the other components of the mixture are also stable. Since decomposition products usually have a different

TABLE III.3

Compounds	Solvent		Support	Operating distribution ratio R_f or p, estimated
	Stationary	Mobile		
Fatty acid, methyl ester C_6, C_8, C_{10}, C_{12}, C_{14}	Isooctane	82% Ethanol–water	Siliconized firebrick	0.67, 0.50, 0.3, 0.2, 0.14
Fatty acids C_9–C_{19}	Paraffin	35–75% Acetone–water	Siliconized Celite	0.5→0.05
Benzene hexachloride isomers	Nitromethane	Hexane	Silicic acid	0.7→0.3
Steroid hormones	40–70% Methanol–water	*n*-Hexane	Super Cel	0.3→0.7
Steroid hormones	50% Methanol–water	Benzene–ethyl acetate	Celite	0.5
Bile acids	$CHCl_3$–heptane (9:1)	60% Methanol–water	Siliconized Celite	0.12–0.8
Phenols	6% Hexanol–cyclohexane	0.5 *M* Sodium chloride	Teflon 6 10–80 mesh	0.18–0.49
Vasopressin	2-Butanol–0.1% acetic acid (water)	2-Butanol	Celite	0.08–0.16
Oxytocin	Butanol–benzene–pyridine–0.1% acetic acid	2 *N* Acetic acid	Sephadex G-25	0.1–0.2
Acetamino acids	Chloroform–butanol (99:1)	Water	Silica	0.27–0.5
DNP amino acids	Citrate–phosphate buffer, pH 3	Butanol	Chloroprene	0.08–0.8
DNP amino sugars	Borate buffer, pH 9.9	67% Amyl alcohol–chloroform	Super Cel	0.3–0.7
Antibiotic Canarius	Phosphate–citrate, pH 6	Ethyl acetate–butanol	Super Cel	0.2
S-RNA	*n*-Butanol, acetic acid, tri-*n*-butylamine, water, di-*n*-butyl ether (lower phase)	(upper phase)	Sephadex G-25	0.7–0.9
S-RNA	Isoamyl acetate quaternary amine	NaCl, tris buffer	Chromosorb	0.06–0.2

APPLICATIONS OF PARTITION CHROMATOGRAPHY

Av HETP (mm)	Column performance, stationary phase (% of vol.)	Flow rate (cm/min)	Load (mg/cm^3)	Remarks	References
1.4	19	0.5	0.0005	Skewing evident	(*24*)
1.0–1.5 (est.)	30	—	0.5–1.0	Column reused 4 times	(*25*)
2	16	0.25	12–16	Separation incomplete	(*26*)
0.4–1.0	38	0.06	0.02	80–90% recovery	(*27*)
20	—	—	0.0004	—	(*28*)
0.25–0.5	30	—	1–1.5	—	(*29*)
3.3	25	2.5	0.1	—	(*30*)
2	20	0.09	0.2–0.5	Oxyquinoline treated to remove heavy metals. Separates oxytocin and vasopressin	(*31*)
5	40	0.09	0.7	—	(*32*)
0.02	22	—	0.25–2	—	(*15*)
—	—	—	—	—	(*33*)
4	15	—	0.3	—	(*34*)
20–30	13	0.6	13	Sixfold purification	(*35*)
3	40	0.02	1	Partial separation	(*36*)
3.4	16	1.6	0.5–1.0	Gradient used	(*37, 38*)

partition ratio than the material undergoing decomposition, a potentially good separation can fail because of instability. Although the stability of unwanted materials is difficult to determine, thin-layer or paper chromatographic studies with an aliquot of the material charged to the CCD apparatus and stored under conditions of the run may indicate the formation of decomposition products and whether the separation has been influenced.

In starting with a substance of unknown properties valuable data concerning possible solvent pairs can be obtained by the paper or thin-layer chromatographic methods previously outlined. Other methods of determining the partition ratio have been mentioned.

B. Equipment and Methods of Operation

Three main types of laboratory instruments are available. In the first (classical Craig type) the volume of the lower phase is fixed by the apparatus but the upper-phase volume can be adjusted from about 1/4 to 2 times the lower-phase volume. In operation the cell containing both phases is mixed by rocking about a horizontal axis. After mixing the cell is maintained stationary at a slight angle until the solvents have separated. Then the cell is slowly tilted to a vertical axis allowing the upper phase to decant into a transfer tube. On return to the horizontal position the solvent flows into the next tube. Trains having up to 1000 cells have been made. Since the stationary phase always remains in the original cell, the apparatus can be operated in any way which utilizes transfer of the upper phase. If all of the material is charged into the first cell and the procedure stopped at any time at which the solvent originally present in the first cell is still in the apparatus, this is called a fundamental distribution. The contents of each individual cell are emptied and analyzed. If the volumes of solvent in each tube and at each transfer are exactly identical then the content of each cell is described by the binomial expansion $(p+q)^n = 1$, in which p is the fraction transferred and q is the fraction remaining.

With the fundamental distribution procedure no more transfers can be used in a particular separation than there are tubes in the apparatus. Greater separation may be obtained by using the withdrawl technique. In this procedure the decanted solvent from the last tube of the apparatus is collected as an individual fraction and the process continued until the desired substance has emerged. For an apparatus with r cells and a substance with a distribution of p, the band would be half emerged (peak concentration) from the apparatus after $(r-1)/p$ transfers and 99.7% emerged after $[(r-1)/p] + 3\,[r(1-p)]^{\frac{1}{2}}$ transfers. This procedure permits greater separation since more transfers can be used. It is especially effective if p is less than 0.5. For larger values of p, the band will emerge from the apparatus after fewer transfers. For this condition a recycle procedure can be used. The band containing the desired substance is recycled by flowing the solvent from the last tube of the apparatus to the first tube. This procedure, in theory at least, can be continued until slower moving impurities begin to emerge.

A less common but sometimes convenient method of operation is the "completion of squares" procedure. In this method the solvent feed is discontinued at some point and the transfers continue until no more solvent remains in the train.

A second type of CCD apparatus has been described by Verzele (*60*). This equipment can transfer either upper or lower phases depending upon a preselected program. This allows even greater resolution in a single extraction train and permits the use of systems having partition ratios which would be unsatisfactory in the forward decantation type. This instrument can also be operated at steady state with continuous feeding. Under steady state conditions, the material removed from the train in some unit of time or per transfer must equal the material charged. Since the ratio and sequence, of upper and lower phase transfers can be programmed with this apparatus, it can be used in the preparation of large quantities of material having constant partition properties. This is possible even though the partition ratio changes with concentration since, at steady state each tube has a constant composition. In addition, a much higher concentration of material can be maintained in the apparatus than is possible in the usual decantation type.

A third type of apparatus has recently been described by Post and Craig (*61*). After equilibration and separation of the phases, tipping the cell causes the lower phase to move to an adjacent cell and the upper phase to move to the other adjacent cell. Thus in operation lower phase emerges from one end of the train and upper phase from the other end. This apparatus, while having only half the resolving power per tube, is especially suitable for steady state applications. If relatively large quantities of material are being processed and high resolution is not required, the fact that two transfers are performed for each equilibration results in a more rapid processing of material.

C. Separation Ability

Although partition chromatography has been treated on the plate theory, it was pointed out that the factors contributing to band spread, especially diffusion, might vary with the compounds being separated and the same column would give different plate heights for different compounds. With CCD as long as the distribution coefficient remains independent of concentration and the equilibration and separation steps are adequate all components should give one theoretical plate per transfer. Since the theory requires that each transfer be exactly the same as all other transfers, the equipment must be precision-made and the operation carefully controlled. The requirements of purity for compounds may vary depending on the circumstances. Relatively few transfers can give substantial enrichment and may be adequate under certain circumstances. In addition the conditions used may influence the number of transfers that can be carried out. Factors such as stability, sensitivity of detection, and absolute solubility can influence the operation and the system used.

The relationship between distribution coefficient, number of transfers, and the degree of separation has been developed previously. The following discussion will be the final mention of the influence of these factors in separation. Frequently, the separation ability is considered as a problem of two components present in equal quantities. In actual practice there are usually many components and they are not present in equal amounts, therefore plots of Gaussian curves of equal size may be deceiving.

The major factors which influence CCD operation are contained in Tables III.4–III.6, which have been prepared for five values of the distribution

TABLE III.4

BAND SPREAD OF A SINGLE COMPONENT IN AN IDEAL CCD, PERCENTAGE OF TOTAL TUBES CONTAINING 95% AND 99.7% OF MATERIAL CHARGED[a]

	Distribution coefficient, G									
	0.1, 10		0.2, 5.0		0.4, 2.5		0.8, 1.25		1.0	
No. of transfers	95%	99.7%	95%	99.7%	95%	99.7%	95%	99.7%	95%	99.7%
25	22.4	34.4	29.9	44.8	36.0	54.0	39.9	59.8	40.0	60.0
50	16.2	24.2	21.1	31.7	25.6	38.4	28.0	42.0	28.4	42.6
100	11.5	17.2	15.0	22.4	18.0	27.0	19.9	29.9	20.0	30.0
200	8.1	12.1	10.6	15.9	12.8	19.2	14.0	21.0	14.2	21.3
400	5.8	8.6	7.5	11.2	9.0	13.5	10.0	15.0	10.0	15.0
800	4.1	6.1	5.3	7.9	6.4	9.6	7.0	10.5	7.1	10.7
1600	2.9	4.3	3.7	5.6	4.5	6.8	5.0	7.5	5.0	7.5

[a] Calculated using $\sigma = (nG)^{\frac{1}{2}}/(G+1)$, $4\sigma = 95\%$, $6\sigma = 99.7\%$.

coefficient and apply to their reciprocals as well. The distribution coefficient G, is taken as the operating condition and is assumed to be constant. The values are those of a Gaussian distribution and represent a good approximation as long as the number of transfers is greater than 25 for a fundamental or completion of squares type distribution and greater than 100 for a withdrawal type distribution.

The values contained in Table III.4 which shows the percentage of all the

TABLE III.5

CONCENTRATION IN SELECTED TUBES OF A SINGLE COMPONENT IN AN IDEAL CCD[a]

No. of transfers	Distribution coefficient, G																			
	0.1, 10				0.2, 5				0.4, 2.5				0.8, 1.25				1.0			
	C_{max}	$\pm 1\sigma$	$\pm 2\sigma$	$\pm 3\sigma$	C_{max}	$\pm 1\sigma$	$\pm 2\sigma$	$\pm 3\sigma$	C_{max}	$\pm 1\sigma$	$\pm 2\sigma$	$\pm 3\sigma$	C_{max}	$\pm 1\sigma$	$\pm 2\sigma$	$\pm 3\sigma$	C_{max}	$\pm 1\sigma$	$\pm 2\sigma$	$\pm 3\sigma$
25	27.8	16.9	3.7	0.3	21.1	12.8	2.8	0.2	17.8	10.8	2.4	0.2	16.0	9.7	2.2	0.2	15.9	9.6	2.2	0.2
50	19.7	11.9	2.7	0.2	15.0	9.1	2.0	0.2	12.6	7.6	1.7	0.1	11.4	6.9	1.5	0.1	11.3	6.9	1.5	0.1
100	13.2	8.0	1.8	0.1	10.7	6.5	1.4	0.1	8.9	5.4	1.2	0.1	8.1	4.9	1.1	0.1	8.0	4.8	1.1	0.1
200	9.9	6.0	1.3	0.1	7.6	4.6	1.0	0.1	6.3	3.8	0.8	0.1	5.7	3.5	0.8	0.1	5.7	3.5	0.8	0.1
400	7.0	4.2	0.9	0.1	5.4	3.3	0.7	0.1	4.5	2.7	0.6	0.1	4.1	2.5	0.6	—	4.0	2.4	0.6	—
800	4.9	3.0	0.7	0.1	3.8	2.3	0.5	—	3.2	1.9	0.4	—	2.9	1.8	0.4	—	2.8	1.7	0.4	—
1600	3.6	2.1	0.5	—	2.7	1.6	0.4	—	2.3	1.4	0.3	—	2.0	1.2	0.3	—	2.0	1.2	0.3	—

[a] Approx. conc. of tube at C_{max}, $C_{max} \pm 1\sigma$, $\pm 2\sigma$, $\pm 3\sigma$. Concentration $= C_{feed} \times \%$. C_{max} calc. from $G + 1/(2\pi n G)^{\frac{1}{2}}$, values rounded to 0.1%; this represents a maximum value based on the assumption that the tube indicated occurs at the precise point of C_{max}, etc.

TABLE III.6

DILUTION OF BAND CONTAINING $\pm 1\sigma$, 68%; $\pm 2\sigma$, 95.4%; AND $\pm 3\sigma$, 99.7% OF A SINGLE COMPONENT IN AN IDEAL CCD[a]

No. of transfers	Distribution coefficient, G														
	0.1, 10.0			0.2, 5.0			0.4, 2.5			0.8, 1.25			1.0		
	± 1[b]	$\pm 2\sigma$	$\pm 3\sigma$	$\pm 1\sigma$	$\pm 2\sigma$	$\pm 3\sigma$	$\pm 1\sigma$	$\pm 2\sigma$	$\pm 3\sigma$	$\pm 1\sigma$	$\pm 2\sigma$	$\pm 3\sigma$	$\pm 1\sigma$	$\pm 2\sigma$	$\pm 3\sigma$
25	23.7	16.5	11.5	18.2	12.8	8.9	15.1	10.6	7.4	13.6	9.5	6.7	13.6	9.5	6.7
50	16.9	11.8	8.3	12.9	9.0	6.4	10.6	7.5	5.3	9.6	6.8	4.8	9.5	6.8	4.8
100	11.8	8.3	5.9	9.1	6.4	4.5	7.6	5.3	3.7	6.8	4.8	3.4	6.8	4.8	3.4
200	8.4	5.9	4.2	6.4	4.5	3.2	5.3	3.8	2.6	4.8	3.4	2.4	4.7	3.4	2.4
400	5.9	4.2	2.9	4.5	3.2	2.7	3.8	2.6	1.9	3.4	2.4	1.7	3.4	2.4	1.7
800	4.2	2.9	2.1	3.2	2.7	1.6	2.6	1.9	1.3	2.4	1.7	1.2	2.4	1.7	1.2
1600	2.9	2.1	1.4	2.3	1.6	1.2	1.9	1.3	0.9	1.7	1.2	0.9	1.7	1.2	0.9

[a] Concentration $= C_{feed} \times \%$.
[b] Calculated assuming no fractional tubes are included in band.

tubes of the distribution occupied by a zone having a particular distribution coefficient have been plotted around the peak point for that distribution in Fig. III.7. The degree of separation for compounds having a β of 2 at the low end and of 1.25 at the high end can be seen as a function of both 4σ and 6σ band spread.

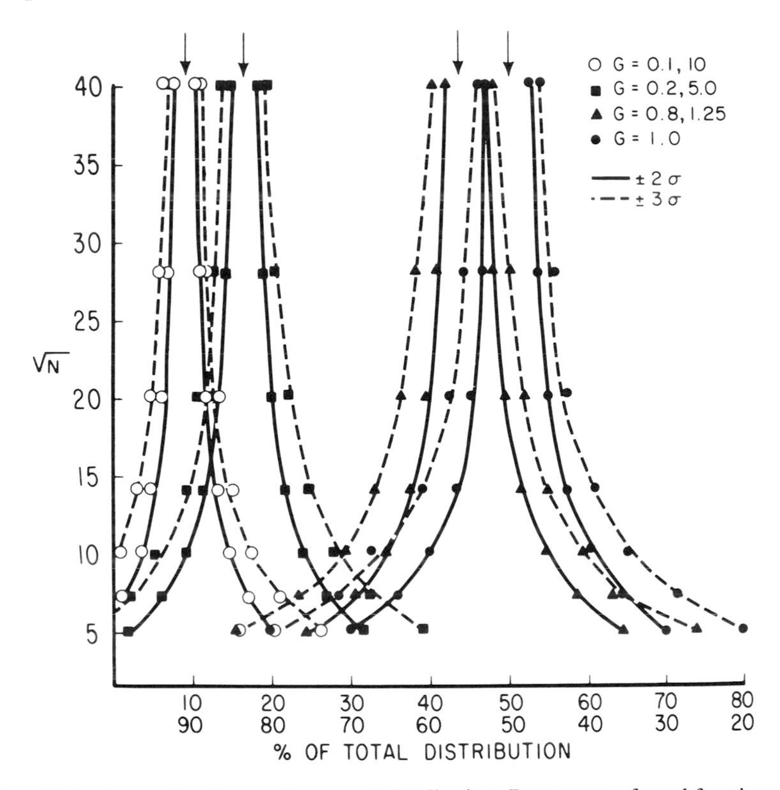

FIG. III.7. Resolution in countercurrent distribution. Percentage of total fractionation occupied by 4σ, 95.4%, and 6σ, 99.7%, of a single substance as a function of distribution coefficient and the number of transfers.

Since the expense of the equipment and the expenditure of time, materials, and effort for a CCD separation are directly linked to the number of transfers, the probable usefulness of CCD equipment of various sizes and tube numbers is influenced by a number of factors. Usually a CCD apparatus can perform about 200 transfers per 24-hr day. Allowing for setup and cleanup time a run of 400–600 transfers requires about one working week. With a 1000-tube machine the solvent front would not reach the end of the train. On the other hand with the withdrawal procedure a substance with a distribution coefficient of 1 would travel through a 200-tube machine and a substance with a coefficient of 0.3, through a 100-tube machine in this time. Withdrawal operation has the advantage that an analysis can be carried out on the withdrawn samples without opening the tubes of the machine and with no loss of machine operating time. On the other hand once material has emerged no further fractionation can be obtained. Since the difference in purification obtained with a G of 1 and that with a G of 0.3 at 400–500 transfers is not great (Fig. III.7), equipment with more than 200 tubes is rarely needed except for exotic separations. If the major purpose of the run is analytical, this can usually be carried out with a relatively small number of transfers as discussed below.

For preparative purposes some loss in yield can usually be taken to improve purity. Examination of Fig. III.7 indicates that a considerably higher purity is obtained by collecting a band containing only 95% of a single component rather than a band containing 99.7%. If a subsequent purification step, such as crystallization, is anticipated, the use of larger trains and more transfers may be contraindicated. Table III.6 indicates the effect of increasing transfer numbers and the yield taken on the total dilution of the fraction collected. Thus, a fraction containing 95% of a single component from a 400-transfer run has a concentration of about 1/40 that of the feed concentration.

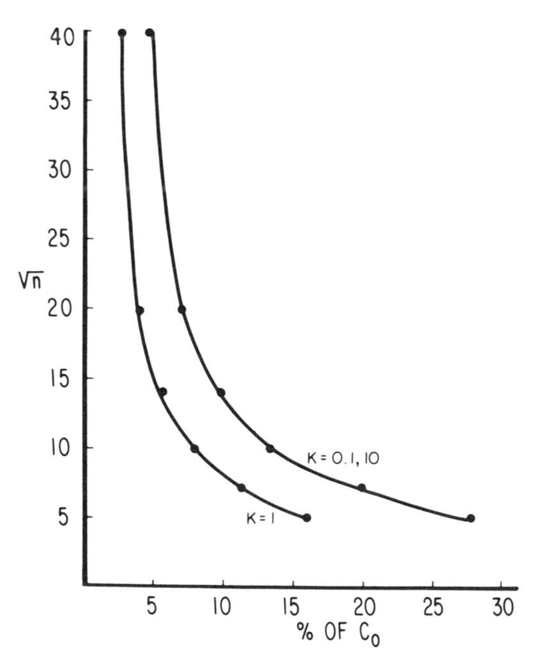

FIG. III.8. Peak tube concentration in countercurrent distribution. Influence of partition ratio and number of transfers.

A fraction containing 99.7% has a concentration of 1/60 that of the feed concentration. The last 4.7% is very dilute and in addition may contain more impurity than the major portion. Since removal of solvent can be a major problem with heat-sensitive materials, this factor should be considered in planning a CCD fractionation.

The nature of the analytical problems encountered in carrying out a CCD fractionation is indicated by Table III.5 and Fig. III.8. The concentrations given in Table III.5 are those obtained if all the material is charged to tube 0. The size of the distribution train may be important in preventing excessive loss of the substance in the analysis itself. For example, if a chemical assay having good accuracy at 10 μg is used and the starting concentration of the desired compound is 2 mg/ml, analysis of the 2σ band spread would require about a 1-ml sample. This is 20% of the material charged to an apparatus with 5-ml phases.

Frequently the charge may be distributed over several tubes at the start of the run. After completion of more than 100 transfers this does not influence the shape of the distribution appreciably if up to about 5 tubes are filled. The resulting distribution pattern represents a composite of a series of Gaussian curves in sequence. Placing the charge in the first 5 tubes would thus cause the peak to appear as though the distribution started in tube 2. The absolute concentration would be about 5 times that obtained if the same concentration were placed in a single tube.

A small number of transfers can be quite useful in determining the purity of a substance of relatively high purity. If a system is available in which the distribution ratio is about 1, a 7–10 tube distribution carried out in small centrifuge tubes or similar equipment can be analyzed for purity. After completion of the distribution under conditions which ensure no loss of solvent due to evaporation and in which special care is taken to ensure complete transfer of the mobile phase after each equilibration, the amount of substance present in one or more tubes is accurately determined. In a distribution using only 8 tubes the presence of 3% of an impurity with a separation factor of 4 can be detected.

The use of a Craig type extraction train with 50 tubes for the determination of the purity of radiochemicals has been described by Sheps *et al.* (*12*).

D. Information Needed for CCD Operation

The effectiveness of CCD in a particular problem can be assured and the most effective use of equipment and time obtained by applying certain preliminary tests. A few small-scale probe type runs should be made to determine problems which may be encountered. From these a good approximation of the partition ratio of the desired compound should be obtained.

The system can be observed for emulsification and phase volume changes which can develop during the separation. If more than one system can be used, factors such as recovery from the solvent, probability of constant partition ratio, and β can be evaluated.

If a buffered system is used for organic acids or bases buffering ability should be adequate to accommodate the change in concentration which will be obtained during the separation.

An important consideration may be the ease of analysis. Certain solvents or buffers interfere with spectrophotometric measurements, color reactions, or bioassays, whereas others may be used without removal. Unless some adequate detection method is available, a CCD run should not be attempted.

Although volatile solvents are desirable, generally, some solvents are highly inflammable and may be undesirable for safety reasons. Potentially explosive substances, e.g., perchlorates, nitrates, and picrates, should be avoided if possible.

E. Factors Affecting the Separation Factor, β

Although it was stated earlier that selectivity, β, usually does not change markedly with the use of a different solvent system, a relatively small change can be important in determining the usefulness of a system for CCD. For some systems the relationship of the partition ratio was shown to be approximated by the following expressions:

$$\log K = a \log K_1 + b \tag{III.10}$$

thus

$$\log \beta = a \log \beta_1 \tag{III.11}$$

For two systems, as the constant a approaches 1, differences in β disappear. Likewise as β approaches 1, differences between two systems become less. For CCD, however, even slight differences in β should be exploited. If two substances have a β value of 1.25, a second system with an a constant of 1.25–2.0 would give β values of 1.1–1.6. Referring to Fig. III.7, it can be seen that a separation of compounds with a β of 1.1 is very difficult, whereas if the value is 1.6 a good separation is usually feasible.

For ternary systems having one phase of nearly identical composition the partition ratio has been stated to be a function of the mole fraction of the remaining two solvents

$$\log K = x \log K_a + y \log K_b \tag{III.12}$$

For systems obeying this relationship a change in proportions of the solvents causes a change in β for two components. The extent of this change is limited by the requirement that the most desirable partition ratio for CCD is nearly 1. For such systems, however, it may be advantageous to deliberately choose a system which does not have the ideal partition ratio if a resulting increase in the separation factor is obtained.

The resolution power of CCD (as well as partition chromatography) for weakly ionized substances having different p*K* values is obviously dependent on the pH of the separation. Since the partition ratio is an exponential function of the pH in the region in which part of the substance is ionized, a large change in β may be obtained within a spacing of 1 or 2 pH units. If the p*K* has been previously determined by chromatography as outlined in Chapter II, probable solvents and pH ranges will be indicated from this information.

For fully ionized acids or bases, separating efficiency can be influenced by the counterions used in the system. Although the change in β obtained with multiply charged bases and various anions is readily apparent from Table II.7, the use of multicharged carrier ions can lead to skewing and loss of resolution. Counterions should be present in relatively large excess to ensure that all of the material being distributed remains as one species. Under these circumstances, if the counterion is singly charged only one species would be distributed regardless of the concentration. Hausmann and Craig (*62*) have reported a case in which failure to remove sulfate in a CCD of a hydrochloride salt led to severe tailing of the peak due to sulfate contamination after even 1000 transfers.

If a nonspecific test is available which locates the desired compound and associated impurities on paper or thin-layer chromatograms, the above factors can be evaluated using the system under study. These may provide a quick method of determining whether the β is actually affected by solvent ratios, pH, and the system employed. The chromatographic system can also be used to evaluate any small-scale probe runs which may be carried out.

The main objective of the experimental approach outlined above is to increase the effectiveness of the CCD itself. Lack of success of CCD separations is usually due to failure to carry out sufficient preliminary tests.

F. Operating Conditions and Variables

If an appropriate system has been found, there may still be different methods of operation depending on the properties of the compounds being fractionated and on the apparatus available. If a large amount of substance is available, it may be charged into a single tube (with possible disturbance of the phase ratios) or spread through several tubes of the apparatus. The latter procedure may be indicated if it is anticipated that the run will require many transfers and if a steady state apparatus is not available or cannot be used for the separation. The effective distribution ratio may be varied by changing ratios of the phases employed. This type of change will affect the number of transfers required. The time course may dictate the desirability of using a particular procedure. Since about 200 transfers are obtained per 24-hr day runs of over 1000 transfers may necessitate operation during 2 weeks. Frequency of sampling for analysis may be dependent on the speed of analysis and the availability of equipment at a particular time.

The run may be planned to deliberately sacrifice yield if starting material is available in abundant supply. On the other hand if material is in limited supply, it may be desirable to choose a condition with high or low distribution ratios since a smaller percentage of the total sample will be consumed by the analysis. Low sensitivity of the assay method may also indicate a preferred operation technique.

The effect of changing the distribution ratio is illustrated in the Fig. III.9. This figure depicts the separation of three compounds having partition ratios of 0.5, 1, and 1.5 in which it has been assumed that the substance B is the desired compound. Solvent ratios of 0.4 and 2.0 were used and the run was assumed to have been carried out by the withdrawal technique to the point at which 99.7% of B has just emerged from a 100-tube apparatus. By examination of Fig. III.9 it is readily apparent that under the second condition only a little over 50% of B could be obtained pure whereas the first condition would have given a high yield of pure B. In addition, the concentration of B in this procedure is higher and less solvent is needed. Longer operating time is required and if the substances being separated are present only in small quantities the overlap depicted may not be serious.

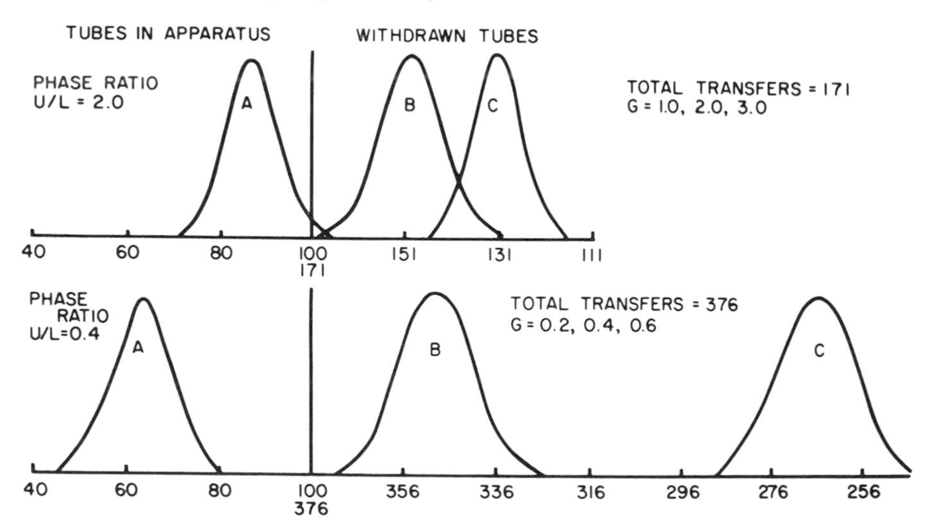

FIG. III.9. Typical separation by countercurrent distribution. Effect of mobile-phase volume; 100-tube decantation-type apparatus. Partition ratios: compound A, 0.5; B, 10. C, 1.5.

G. Applications of CCD

Although CCD would appear to be applicable in any separation of stable compounds, experimental difficulties may cause the method to fail or perform badly. With very polar or nonpolar substances satisfactory two-phase systems may not be available. Problems also occur when the strict requirement for constant partition ratio, solvent composition, and lack of emulsification are not obtained.

1. Nonconstant Partition Ratio

At equilibrium the "activity" of the distributed material is equal in the two phases. Since any system which deviates from Raoult's law has a non-linear partition ratio as the concentration changes, it may be surprising that CCD can be used at all. However, it is frequently observed that the partition ratio is constant or nearly so at low concentration. Occasionally impurities influence the partition ratio by complexing with the desired substance or by changing the nature of the solvent. Such complexing reactions may be used in devising systems having constant partition ratios. The formation of hydrogen-bonded dimers involving the –COOH group of fatty acids usually causes a nonlinear partition ratio. Ahrens and Craig (*63*) showed that solvent mixtures which contain appreciable quantities of acetic acid are useful for the CCD of fatty acids. Since a large excess of acetic acid is present it is probable that the species being distributed is a complex of the fatty acid and acetic acid. For ionized materials the desirability of using monofunctional counterions has been mentioned.

Distribution curves which would be obtained for compounds which form dissociable complexes have been calculated by Bethune and Kegeles (*64*). Distribution curves were calculated for cases in which one constituent dimerizes, trimerizes, or reacts with a second compound. Dimerizing compounds produce skewed curves with one boundary sharp and one trailing. If the partition ratios of reacted species and the desired compound are sufficiently different and the complex readily dissociates, separation may be obtained at some point in the distribution.

Experimentally obtained and computer-generated curves for the cyclic decapeptide antibiotic, tyrocidine, are in good agreement (*65*). In the system used (chloroform, methanol, 0.01 *N* HCl; 2:2:1) the partition ratio increases with concentration, probably due to complex formation between two or more peptide molecules. Natural tyrocidine is a mixture of three compounds having the following structures:

$$\begin{array}{l} \text{L-Val} \rightarrow \text{L-Orn} \rightarrow \text{L-Leu} \rightarrow \text{D-Phe} \rightarrow \text{L-Pro} \rightarrow \text{L-X} \\ \uparrow \qquad\qquad\qquad\qquad\qquad\qquad\qquad\qquad\qquad\qquad \downarrow \\ \text{L-Tyr} \leftarrow \text{L-GluN} \leftarrow \text{L-AspN} \leftarrow \text{D-Y} \leftarrow\!\!-\!\!-\!\!- \end{array}$$

Tyrocidine	Residue X	Residue Y	Partition ratio
A	Phe	Phe	1.1
B	Try	Phe	2.0
C	Try	Try	3.2

The approximate partition ratios at high concentration, 5 mg/ml, are indicated in the above table. At low concentration all the partition ratios are much lower and more nearly identical. If complexes of the type A–A, A–B, A–C, etc., can form, it would be expected that the separation would be quite difficult. In addition, for very large numbers of transfers or for a CCD separation carried out at low concentration, some separation may be obtained which is then lost because the boundaries trail more and more into the next zone as the distribution continues. Both of these situations do, in fact, occur. In spite of these severe difficulties it was demonstrated that at certain feed concentrations and at an intermediate stage in the CCD operation good purification of the various types could be obtained. This requires many more transfers than would be expected from the partition ratios of the pure compounds.

For polar compounds it is usually necessary to use a solvent system containing water. Many such systems are influenced by the presence of compounds in the sample which hydrate and thus disturb the phase ratios. Occasionally emulsions develop during the course of a CCD run even though not present at the beginning. In a decantation type apparatus retention of the mobile phase due to loss of the immobile phase can cause serious loss of the efficiency of the separation.

Several extensive reviews of the application of CCD are available (*9*, *66–70*).

A number of examples have been chosen to illustrate the use of the method with compounds of differing polarity and with different types of equipment and at different scales of operation.

2. *Preparative-Scale Applications—Low Transfer Numbers*

Work-up of a crude concentrate containing one or more active principles by countercurrent extraction may offer advantages over chromatography in that the material is always recoverable. No irreversible or unexpected adsorption can occur. Also the degree of resolution can be tailored to the needs of the problem. With secondary use of thin-layer or paper chromatography in following the CCD, a completely thorough and unequivocal isolation can be performed. Sufficient separation may be obtained in relatively few transfers and special equipment is not needed.

Although with sufficient study and increased knowledge of the properties of an unknown substance, chromatographic procedures are usually more useful and practical for the isolation of a single component, the CCD approach is usually more certain of achieving an initial isolation.

A. VERATRUM ALKALOIDS. The above principles are illustrated in the isolation of *Veratrum* alkaloids (*71*, *72*). In these studies pure material was not obtained directly from the CCD but fractions were sufficiently enriched to yield crystalline, characterizable products. A crude alkaloid extract from *Veratrum* album was crystallized from ether and yielded a crystalline fraction and a supernatant fraction. Using a system of benzene–2 *M* acetate pH 5.5 buffer, a distribution was run in separatory funnels. The ether crystalline product, 39 g, was carried through 24 transfers using 800 ml of each phase. After combining the upper and lower phases of each tube three fractions, 0–9, 10–16, and 17–24 were crystallized, yielding veratetrine, germitetrine, and proveratrine, respectively. The total yield of crystalline products was about 50% of the starting weight. The ether mother liquor which contained the same three alkaloids as well as additional compounds was distributed in 18 plates using a charge of 51 g and 900–ml phase volume. Fractions 7–18 contained 28 g and yielded an additional large quantity of germitetrine. Fractions 0–6 were combined and redistributed in the same volume of a system benzene–2 *M* acetate, pH 6.5, using 30 plates. This gave a higher partition ratio. Fractions 10–20 (total weight 3.1 g) yielded desacetylproto-veratrine. Fractions 0–7 contained 12.7 g of solid which yielded 3.7 g of an impure crystalline product. This was distributed by the withdrawal technique using chloroform–2 *M* acetate, pH 4.4, and 140 transfers. Effluent tubes 85–130 were combined and yielded 840 mg of crystalline neogermbudine. From tubes 36–60 (1.35 g) crystalline desacetylneoprotoveratrine was obtained.

Thus a total of seven compounds were obtained in sufficient quantities to determine chemical and biological properties. It should be pointed out that in this case the CCD did not yield "pure" material or any valid information concerning purity of the fractions.

B. CORYNANTHEINE. Literature reports concerning the presence of an aliphatic double bond in this yohimbine alkaloid were inconsistent. Dihydrocorynantheine and corynantheine readily cocrystallize and repeated chromatography over alumina was reported not to yield improved product. Crude corynantheine can be separated into two fractions using CCD (*73*). A 40-transfer CCD gave the mass distribution illustrated in Fig. III.10. The double bond content of the starting material was 66%. The presence of a less polar impurity (K about 1.1, $\beta = 1.75$) is indicated. Although separation was incomplete, combining fractions 21–30 and crystallization yielded pure corynantheine with

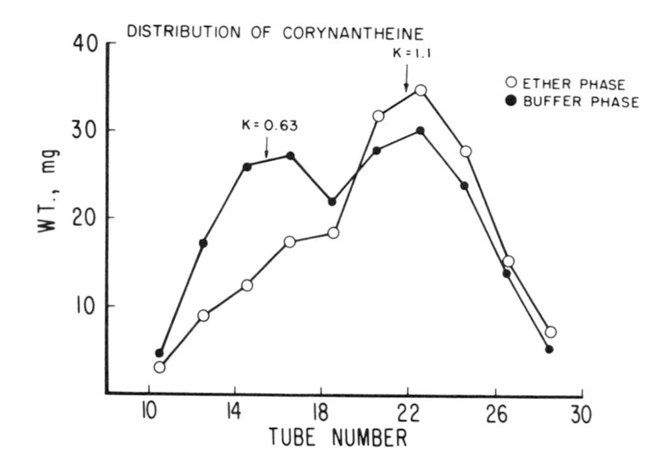

FIG. III.10. Separation of corynantheine and dihydrocorynantheine by countercurrent distribution. The system was ether, 0.2 *N* sodium citrate buffer, pH 4.1; 25 ml each phase, 32 vessels, 40 transfers; charge, 400 mg corynantheine with 66% double bond content.

the expected hydrogen uptake for one double bond. A preparative run was then carried out using 1-liter phases and 20 g of crude material. The separation was obtained by extracting 1 liter of ether with 21 portions of buffer. The buffer extracts were then successively extracted with 19 portions of ether. The ether extracts yielded 10.9 g of solid which was crystallized to yield 8.5 g of pure corynantheine.

In this case the nearly constant partition ratio observed in the latter tubes of the CCD is indicative of high purity.

3. High Transfer CCD in the Preparation of Pure Compounds

Countercurrent distribution is especially valuable in the preparation and characterization of biologically active compounds. The procedure has been applied to the separation of lipids, steroids, peptides, nucleic acids, antibiotics and alkaloids. The work of Craig in the preparation of pure peptides is noteworthy. The method is of greatest value for compounds which do not crystallize and for which tests for homogeneity are difficult to apply.

A. OXYTOCIN. Although there are many examples of the use of CCD in the literature the following are illustrative. During the isolation and synthesis of the peptide hormone, oxytocin, CCD was used as the major fractionation method. Several solvent systems were found useful in these studies. A synthetic derivative, containing D-cysteine in the amine terminal position, was observed to have significant biological activity when prepared by one method of synthesis (*74*).

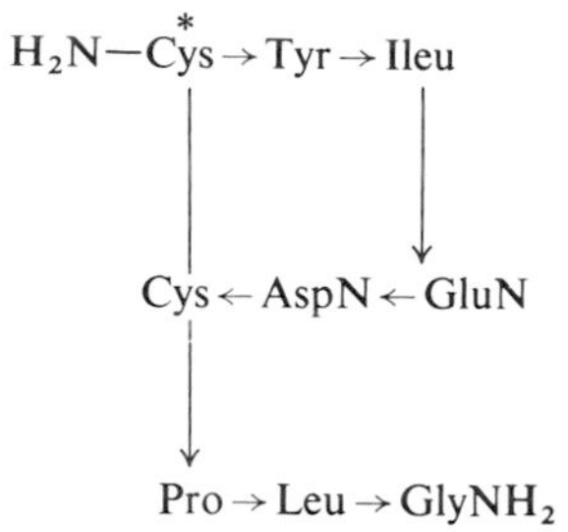

$$\begin{array}{l} H_2N{-}\overset{*}{Cys} \rightarrow Tyr \rightarrow Ileu \\ \qquad | \qquad\qquad\quad \downarrow \\ \quad Cys \leftarrow AspN \leftarrow GluN \\ \qquad \downarrow \\ \quad Pro \rightarrow Leu \rightarrow GlyNH_2 \end{array}$$

Oxytocin

However, the compound prepared in another laboratory by another method of synthesis was reported to be nearly devoid of activity. Racemization of the cysteine during coupling (a protected *p*-nitrophenyl ester was used) would produce some normal oxytocin. By adding tritiated oxytocin (synthesized by another route) to the CCD run, evidence was obtained which indicated that the bioactivity was associated with a minor fraction of the total. The CCD run was carried out in the system, 1-butanol–propanol–0.05% acetic acid (3:2:5) at 4°C. A recycle procedure was used. After 800 transfers the peak of radioactivity indicated a distribution coefficient of 0.48, whereas the peptide peak (Folin-Lowry determination) corresponded to a distribution of 0.52. After an additional 400 transfers two fractions were made comprising 2.5 σ before the peptide peak and 2.5 σ after the peak. One fraction contained 95% of the biological activity and 99% of the radioactivity. The other fraction, probably the D isomer, was nearly inactive. Although in this case the addition of radioactive oxytocin simplified the analysis it should be pointed out that the use of bioactivity alone would indicate divergence from the main peptide peak. Thus, the D and L isomers have a β of 1.04 in the system used and were separated sufficiently to indicate that the D isomer is inactive or nearly so.

B. NUCLEIC ACIDS. Countercurrent distribution may be the preferred method of separation of large molecules for which slow diffusion and the tendency for irreversible adsorption make chromatography difficult. The elegant work of Holly (Apgar *et al.*, *11*) in the fractionation of transfer RNA illustrates this point. The mixture of nucleic acids poses a difficult separation problem which has been only partly solved. Even if a group separation method could be applied which yielded pure t-RNA, the resulting mixture would contain at least 20 closely related compounds in varying proportions. It is unlikely to expect that a single fractionation would yield a homogenous product. In addition assays for specific activity are cumbersome and slow. Traces of ribonuclease from any contamination by microorganisms (buffer, glassware, and perspiration) can cause serious losses in dilute solutions.The compounds are quite polar and require systems which contain appreciable fractions of water in each phase. However, in the system containing pH 6 phosphate buffer, 2-propanol and formamide partition ratios of the t-RNA's are 0.12 to 4.25, a range of 36-fold. Although certain t-RNA's are very similar, Holly and co-workers were successful in fractionating specific compounds. In a typical procedure fractions obtained from a preliminary 200-transfer distribution were redistributed for 800–1200 transfers in the same or a similar system. This procedure yielded material of 40–60% purity based on amino acid acceptor activity.

4. *Steady State Distribution*

The use of steady state systems in which material is continuously fed into a fractionating train has been somewhat limited, although larger quantities of material can be prepared by this procedure. The resolving power per stage is much higher than that obtained in a decantation type train (*75*). Maximum separation is obtained if the feed is introduced in the middle of the train and the product of the distribution ratios is 1 ($G_a G_b = 1$). Volume ratios should be adjusted to meet this requirement. Thus, two compounds having $K = 1$ and $K = 4$ are separated at maximum efficiency if the phase volumes are adjusted so that $G = \frac{1}{2}$ and 2. Many aspects of the procedure are similar to distillation and one method, using vessels of different size to approximate reflux, has been described (*76*).

A. STREPTOMYCIN A AND STREPTOMYCIN B. The procedure is best suited for the separation of a binary mixture or for removing impurities which are all either more or less polar than the desired substance. The separation of the binary mixture of streptomycin A, a trisaccharide antibiotic, and streptomycin B, a tetrasaccharide, illustrates the method. Both antibiotics contain two strongly basic guanido groups and one weakly basic amino group. Partition ratios into polar organic solvents are poor unless a strongly solvent-seeking anion is used. O'Keeffe *et al.* (*77*) studied the partition of streptomycin in a system containing lauric acid (15% in the solvent phase) and Pentasol, a commercial mixture of amyl alcohols. The partition is pH-dependent in the region of partial ionization of the weakly basic amino group and diminishes rapidly at pH values lower than about pH 6.0 due to the decreased ionization of the carrier acid. Using phosphate-borate buffer at pH 7.15, the partition ratio increases with decreasing concentration but it is reasonably constant at a streptomycin content of 6–10 mg/ml. Under these conditions the partition ratio for streptomycin A is 2.2 and that for streptomycin B, with its additional polar moiety, is 0.63. The separation factor, β, is thus about 3.5 and the optimum ratio of the phases $[(2.2 \times 0.63)^{\frac{1}{2}}]$ is 1.18. The separation was carried out using 11-liter separatory funnels containing 450 ml of lower phase and 495 ml of upper phase. After each equilibration emerging upper and lower phases were removed and solid streptomycin mixture (66% A and 33% B) was added to the center vessel. Since the partition ratio had been shown to be concentration-dependent, the amount added with each cycle should be such that the concentration remains nearly in the optimum region throughout the whole train. If one component is to be removed from each end of the train it is clearly not possible to maintain conditions of constant partition ratio for either component. It would be expected that the substance with the lower partition ratio would "leak" out the solvent end of the train, whereas the higher partition ratio material would not extend as far to the buffer side as anticipated.

In the above case the feed added amounted to about 5 mg/ml. Steady state operation was nearly achieved after 18 additions at a concentration of about 10 mg/ml in the center vessel. The solvent effluent contained streptomycin A of 99% purity and the buffer effluent contained streptomycin B in 85% purity.

Recently Post and Craig (*61*) have described an automated instrument which carries out the type of distribution described above. The efficiency of each cell was shown to be greater than 80%.

A different distribution pattern is obtained with the Verzele type apparatus which transfers either upper or lower phase (but not both) after each equilibration. The use of this equipment and procedures for rapid achievement of steady state conditions are discussed by Alderweireldt (*78*). The efficiency of transfer was shown to be 75–80% with binary mixtures of cresols. Under these circumstances the Craig-Post and the Verzele apparatuses are capable of separating equal amounts of substances having β values in the range of about 1.2 per 100 tubes. The preparative use of the latter equipment was demonstrated by preparing 12 g of cohumulone in 1 week (*78*).

REFERENCES

1. Pierotti, G. J., Deal, C. H., and Derr, E. L., *Ind. Eng. Chem.* **51**, 95–102 (1959).
2. Collander, R., *Acta Chem. Scand.* **4**, 1085–1098 (1950).
3. Collander, R., *Acta Chem. Scand.* **5**, 774–790 (1951).
4. Bowman, M. C., and Beroza, M., *Anal. Chem.* **38**, 1544–1549 (1966).
5. Perisho, C. R., Rohrer, A., and Thoma, J. A., *Anal. Chem.* **39**, 737–744 (1967).
6. Soczewinski, E., Waksmundski, A., and Maciejewics, W., *Anal. Chem.* **36**, 1903–1905 (1964).
7. Soczewinski, E., *Nature* **191**, 68–69 (1961).
8. Wolf, F. J., U.S. Patent No. 3,125,566, (1964).
9. King, T. P., and Craig, L. C., *Methods Biochem. Analy.* **10**, 201–228 (1962).
10. Carpenter, F. H., McGregor, W. H., and Close, J. A., *J. Am. Chem. Soc.* **81**, 849–855 (1959).

11. Apgar, J., Holly, R. W., and Merrell, S. H., *J. Biol. Chem.* **237**, 796–802 (1962).
12. Sheps, M. C., Purdy, R. H., Engel, L. L., and Oncley, J. L., *J. Biol. Chem.* **235**, 3042–3048 (1960).
13. Denkewalter, R. G., Hughey, G. B., and Kutosh, S., U.S. Patent No. 2,635,985, (1953).
14. Rowley, D., Steiner, H., and Zimkin, E., *J. Chem. Soc. Ind.* **65**, 237–240 (1946).
15. Martin, A. J. P., and Synge, R. L. M., *Biochem. J.* **35**, 1358–1368 (1941).
16. van Duin, H., *Nature* **180**, 1473 (1957).
17. Giddings, J. C., *Anal. Chem.* **35**, 2215–2216 (1963); *Anal. Chem.* **35**, 439–449 (1963).
18. Klinkenberg, A., and Sjenitzer, F., *Chem. Eng. Sci.* **6**, 258–270 (1956).
19. Van Deemter, J. J., Zuiderweg, F. J., and Klinkenberg, A., *Chem. Eng. Sci.* **5**, 27–89 (1956).
20. Slobin, L. I., and Carpenter, F. H., *Biochemistry* **2**, 22–28 (1963).
21. Simpson, S. A., Tait, J. F., Wettstein, A., Nehrer, R., Euw, J. V., Schindler, O., and Reichstein, T., *Helv. Chim. Acta* **37**, 1163–1200 (1954).
22. Schmidt-Kastner, G., *Naturwissenschaften* **51**, 38–39 (1964).
23. Zeleznick, L. D., *J. Chromatog.* **14**, 139–141 (1964).
24. VandenHeuvel, F. A., and Sipos, J. C., *J. Chromatog.* **10**, 131–140 (1963).
25. Garton, G. A., and Lough, A. K., *Biochim. Biophys. Acta* **23**, 192–195 (1957).
26. Ramsey, L. L., and Patterson, W. I., *J. Assoc. Offic. Agr. Chemists* **29**, 337–346 (1946).
27. Butt, W. R., Morris, P., Morris, C. J. O. R., and Williams, D. C., *Biochem. J.* **49**, 433–438 (1951).
28. Ayres, P. J., Garrod, O., Simpson, S. A., and Tait, J. F., *Biochem. J.* **65**, 639–646 (1957).
29. Sjovall, J., *Acta Physiol. Scand.* **29**, 1267–1270 (1953).
30. Fritz, J. S., and Hedrick, C. E., *Anal. Chem.* **37**, 1015–1018 (1965).
31. Condliffe, P. G., *J. Biol. Chem.* **216**, 435–464 (1955).
32. Yamashiro, D., *Nature* **201**, 76–77 (1964).
33. Partridge, S. M., and Swain, T., *Nature* **166**, 272–273 (1950).
34. Annison, E. F., and James, A. T., *Biochem. J.* **48**, 477–482 (1951).
35. Argoudelis, A. D., De Boer, C., Eble, T. E., and Herr, R. R., U.S. Patent No. 3,183,154, (1964).
36. Berquist, P. L., Baguley, B. C., Robertson, J. P., and Ralph, R. K., *Biochim. Biophys. Acta* **108**, 531–539 (1965).
37. Kelmers, A. D., *J. Biol. Chem.* **241**, 3540–3545 (1966).
38. Kelmers, A. D., Novelli, G. D., and Stulberg, M. P., *J. Biol. Chem.* **240**, 3979–3983 (1965).
39. Wittenberg, J. B., *Biochem. J.* **65**, 42–45 (1957).
40. Van De Kamer, J. H., Pikaar, N. A., Bolssens-Frankena, A., Couvee-Ploeg, C., and Van Ginkel, L., *Biochem. J.* **61**, 180–186 (1955).
41. Howard, G. A., and Martin, A. J. P., *Biochem. J.* **46**, 532–538 (1950).
42. Green, T., Howitt, F. O., and Preson, R., *Chem. Ind.* (*London*) pp. 591–592 (1955).
43. Badami, R. C., *Chem. Ind.* (*London*) pp. 1920–1921 (1964).
44. Kinnory, D. S., Takeda, Y., and Greenberg, D. M., *J. Biol. Chem.* **212**, 379–383 (1955).
45. Marvel, C. S., and Rands, R. D., Jr., *J. Am. Chem. Soc.* **72**, 2642–2646 (1950).
46. Zbinovsky, V., *Anal. Chem.* **27**, 764–768 (1955).
47. Frankel, E. N., Evans, C. D., Moser, H. A., McConnell, D. G., and Cowan, J. C., *J. Am. Oil Chemists' Soc.* **38**, 130–134 (1961).
48. Corfield, M. C., Fletcher, J. C., and Robson, A., *Chem. Ind.* (*London*) pp. 661–662 (1956).
49. Matheson, N. A., *Biochem. J.* **94**, 513–517 (1965).
50. Chambach, A., and Carpenter, F. H., *J. Biol. Chem.* **235**, 3478–3483 (1960).

51. Porter, R. R., *Biochem. J.* **53**, 320–328 (1953).
52. Peart, W. S., *Biochem. J.* **62**, 520–527 (1956).
53. Porter, R. R., *Biochem. J.* **59**, 405–410 (1955).
54. Hegedüs, H., Tamm, C., and Reichstein, T., *Helv. Chim. Acta* **36**, 357–369 (1953).
55. Bell, D. J., and Palmer, A., *J. Chem. Soc.* pp. 2522–2525 (1949).
56. Hough, L., Jones, J. K. N., and Wadman, W. H., *J. Chem. Soc.* pp. 2511–2516 (1949).
57. Marrain, G. F., and Bauld, W. S., *Biochem. J.* **59**, 136–141 (1955).
58. Bergstrom, S., and Norman, A., *Proc. Soc. Exptl. Biol. Med.* **83**, 71–74 (1953).
59. Lambert, S. M., and Porter, P. E., *Anal. Chem.* **36**, 99–104 (1964).
60. Verzele, M., *Bull. Soc. Chim. Belges* **62**, 619–639 (1953).
61. Post, O., and Craig, L. C., *Anal. Chem.* **35**, 641–646 (1963).
62. Hausmann, W., and Craig, L. C., *J. Am. Chem. Soc.* **76**, 4892–4896 (1954).
63. Ahrens, E. H., Jr., and Craig, L. C., *J. Biol. Chem.* **195**, 299–310 (1952).
64. Bethune, J. L., and Kegeles, G., *J. Phys. Chem.* **65**, 1755–1760 (1961).
65. Williams, R. C., Jr., and Craig, L. C., *Separ. Sci.* **2**, 487–499 (1967).
66. Craig, L. C., and Craig, D., "Technique of Organic Chemistry," 2nd Ed. Vol. 3. Wiley (Interscience), New York, 1956.
67. Casanavi, C. G., *Chromatog. Rev.* **5**, 161–207 (1963).
68. Hecker, E. Z., "Verteilungsverfahren im Laboratorium." Verlag Chemie, Weinheim, 1955.
69. Fleetwood, J. G., *Brit. Med. Bull.* **22**, 127–131 (1966).
70. von Metsch, F. A., *Angew. Chem.* **65**, 586–598 (1953).
71. Klohs, M. W., Draper, M., Keller, F., Koster, S., Malesh, W., and Petracek, F. J., *J. Am. Chem. Soc.* **76**, 1152–1153 (1954).
72. Myers, G. S., Glen, W. L., Morozovitch, P., Barber, R., Papineau-Couture, G., and Grant, G. A., *J. Am. Chem. Soc.* **78**, 1621–1624 (1956).
73. Goutarel, R., Janot, M. M., Mirza, R., and Prelog, V., *Helv. Chim. Acta* **36**, 337–340 (1953).
74. Hope, D. B., Murti, V. V. S., and DuVigneaud, V., *J. Am. Chem. Soc.* **85**, 3686–3688 (1963).
75. Hecker, E. Z., *Z. Naturforsch.* **12b**, 519–527 (1957).
76. Harrison, I. T., *Chem. Ind. (London)* pp. 1526–1527 (1964).
77. O'Keeffe, A. E., Dolliver, M. A., and Stiller, E. T., *J. Am. Chem. Soc.* **71**, 2452–2457 (1949).
78. Alderweireldt, F. C., *Anal. Chem.* **33**, 1920–1924 (1961).

IV *Gel Filtration and Gel-Permeation Chromatography*

Gel filtration and gel-permeation chromatography are separations processes based on molecular size. Separations of the group type (gel filtration) as well as the fractionation type (gel-permeation chromatography) are possible. Ideally there are no adsorption, preferential solubility, ion exchange, or other secondary effects. The process requires an inert solid phase or gel containing an imbibed internal solvent. The solid is usually packed in a column and surrounded by solvent. In operation, the sample is applied and the column developed with the same solvent.

I. MECHANISM

A solute dissolved in the solvent distributes between the external and internal solvent based on its ability to permeate the particle. Thus if two solutes are present, one of which is small (S_1) and permeates the entire internal volume (V_i) while the other is large (S_2) and cannot enter the particle or gel matrix, these materials can be separated by gel filtration. Scheme I depicts the separation which occurs as solvent is passed through the column.

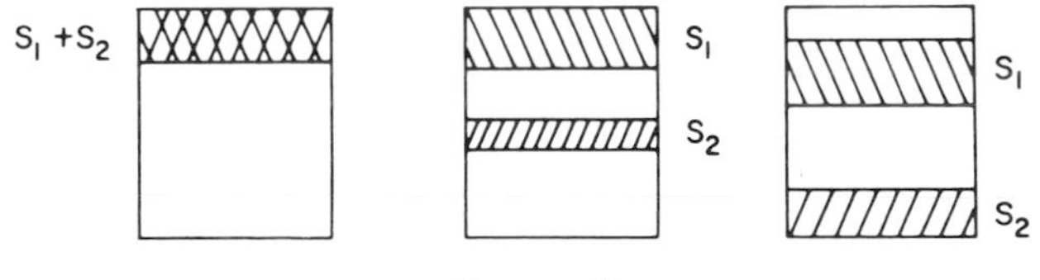

SCHEME I

In the absence of band spread due to diffusional factors and density differences, solute S_2 would emerge from the column in the same volume as that charged, V_s, and would appear at an effluent volume exactly equal to the external volume or void volume (V_0) of the column. In practice the diffusion and mixing which occur in column operation cause the formation of a bell-shaped band with its peak (V_e) occurring at the volume, $V_0 + \frac{1}{2} V_s$. Distortion of the band results from additional disturbing influences, such as convection currents due to temperature changes, nonequilibrated column packing, non-uniform sample, or eluting solvent distribution and density differences between the sample and eluting solvent.

Solute S_1, which completely permeates the particle, would begin to emerge from the column when sufficient eluting solvent to displace the entire liquid volume of the column has been added, i.e., $V_0 + V_i$. For such a solute the concentration at equilibrium is the same in V_0 and V_i. Considering an analogy to partition chromatography the partition ratio, K, between the stationary phase, V_i, and the mobile phase, V_0, would thus have a value of 1. For solutes which do not penetrate the entire internal volume, the partition ratio, K, is that fraction of the internal volume *available* to the solute. It then represents the ratio of the *average* concentration in the internal volume (stationary phase) to the concentration in the external volume (mobile phase). For gel-permeation chromatography this is referred to as K_d. When the sample size is small, it is obtained readily from the following expression:

$$K_d = \frac{V_e - V_0}{V_i} \tag{IV.1}$$

K_d values are usually independent of actual concentration (*1*).

Since the distribution coefficient, G, is ordinarily expressed as the ratio of concentration in the mobile phase to that in the stationary phase for any cross section of the column, its relation to K_d is as follows:

$$G = \frac{1}{K_d} \frac{V_0}{V_i} \tag{IV.2}$$

This relation is analogous to that in partition chromatography and the same relationships of elution volume, band spread, maximum concentration, etc., apply (Chapters I and III). K_d values greater than unity indicate adsorption or some other secondary effect. In the absence of such effects all substances having the same shape and size have the same K_d. Since the volume ratio, V_0/V_i, is usually within the range of 0.7–1.4 for most gels (as long as the particles remain spherical), the distribution coefficient is unfavorable for high resolution. Since G is in the range of 0.7–10, the entire separation must take place in an elution volume corresponding to 0.1–1.4 V_0. Under these circumstances factors other than molecular size which affect K_d may determine the utility.

Based on the assumption that the partition ratio is dependent only on molecular size and shape, gel-permeation chromatography has been utilized as a method of determining molecular weight (*2*). For proteins and other molecules which approximate a spherical shape a linear relationship between the logarithm of the molecular weight and the K_d has been reported (*3*, *4*).

Both retardation and exclusion have been observed to influence K_d. These effects can be eliminated experimentally in the determination of the molecular weight of an unknown substance by carrying out a number of column runs using different solvents or electrolyte backgrounds. For compounds which can be specifically determined in the presence of impurities the gel-permeation chromatography procedure can be used for the estimation of molecular weights even though pure materials are not available. The method is especially useful for impure proteins having enzymatic activity.

A comparison of K_d values by batch and column procedures has been made (*1*, *5*). Generally good agreement was obtained with low and medium porosity gels (Sephadex G-75 and G-100) but the K_d values obtained columnwise are considerably larger than the batchwise determination with very porous gels (Sephadex G-200 and Agarose). Based on the concept that only a certain fraction of the internal volume is available to molecules of a certain size, several mathematical models have been shown to give general agreement with the observed elution volumes (*6*, *7*, *8*).

Similarly, it has also been shown that models based on restricted diffusion within pores of uniform size can account for observed column behavior of large molecules (*5*). Large differences between values for the average pore size of acrylamide gels and agar have been reported (*9*, *10*) for gels are utilizable for the separation of molecules of similar size. Although it is unlikely that completely uniform pore sizes exist and that all of the internal liquid is free to dissolve solutes restricted diffusion probably contributes to the overall K_d. For this reason the best model may include both factors.

II. SUBSTANCES USED FOR GEL SEPARATIONS

A variety of substances have been found useful for the process. Glass particles containing uniform pores have been shown to effect separation of different viral particles from proteins and from each other (*11*). Gels prepared from acrylamide cross-linked with methylenebis acrylamide (*12*, *13*), from dextran polymers cross-linked with epichlorohydrin (*14*), and from starch (*2*), agar (*16*), agarose (*15*), and ion exchange resins (*17*) have been used for aqueous solutions. For lipophilic systems rubber (*18*), polystyrene (*19*), and methylated dextrans (*20*) have been used. Dextran polymers are obtainable in a series of increasing porosity under the name Sephadex. Polyacrylamide gels with graded porosity are also available. The dried gels swell in the solvents employed and the density of polymer within the swollen gel determines the molecular weight range for which the gel is utilizable. Collected data for some commercially available gels are tabulated in Table IV.1.

TABLE IV.1

PROPERTIES OF WATER-SWOLLEN GELS[a]

Designation	Composition of gel phase (vol % dry matter, avg.)	Approximate exclusion m. wt.	
	Biogel		
P-2	40	2,600	
P-4	29.4	4,000	
P-6	21.2	5,000	
P-10	18.2	7,000	
P-30	15.0	50,000	
P-60	12.0	70,000	
P-100	11.8	100,000	
P-150	9.8	150,000	
P-200	6.4	300,000	
P-300	5.4	400,000	
	Sephadex		
G-10	37.5	700	700[b]
G-15	28.6	1,500	1,500
G-25	19.3	5,000	5,000
G-50	10.7	30,000	10,000
G-75	7.9	70,000	50,000
G-100	5.65	150,000	100,000
G-150	3.85	400,000	150,000
G-200	2.9	800,000	200,000

[a] Data taken from manufacturers' information. Biogel resins are cross-linked polyacrylamide and Sephadex resins are cross-linked dextran (see Appendix II).

[b] Exclusion values for dextran polymer.

If hydrophilic gels are placed in aqueous solvents the degree of swelling is reduced and the percentage of dry matter in the swollen gel can be increased. For example, using methanol–water mixtures and Sephadex gels, almost any degree of swelling can be obtained. This is illustrated by data tabulated in Table IV.2.

Similarly the degree of swelling of hydrophobic gels may be influenced by the solvents used (Table IV.3).

TABLE IV.2

SEPHADEX G-25, SOLVENT REGAIN IN METHANOL–WATER

% Methanol	100	95	90	80	70	60	50	40	20	0
Vol change (ml/g)	0.2	0.0	0.1	0.6	1.0	1.7	2.0	2.4	2.8	3.9
% Dry matter in gel phase	59	69	60.5	47	42	33	30	27	24	19.5

TABLE IV.3

SEPHADEX LH-20, SOLVENT REGAIN IN VARIOUS SOLVENTS

Solvent	Solvent regain (ml/g gel)
Dimethylformamide	2.2
Water	2.1
Methanol	1.9
Ethanol	1.8
Chloroform[a]	1.8
n-Butanol	1.6
Dioxane	1.4
Tetrahydrofuran	1.4
Acetone	0.8
Ethyl acetate	0.4
Toluene	0.2

[a] Containing 1% ethanol.

Some relationship exists between the percentage of solid in the gel matrix and the size of molecules excluded. This is shown empirically in Fig. IV.1.

Chemical Properties of Gels

The efficiency of separation obtained under equivalent mechanical conditions (particle size, flow rate, uniformity of packing, etc.) depends on the properties of the gel.

1. Dispersion of Porosity

Gels of completely uniform porosity can be visualized. Such a gel would produce a rather sharp size classification in that substances would either completely permeate the gel or would be excluded. Similarly, the other extreme in which a complete spectrum of pore sizes is present is possible. This would give poor separation since only a small fraction of the total

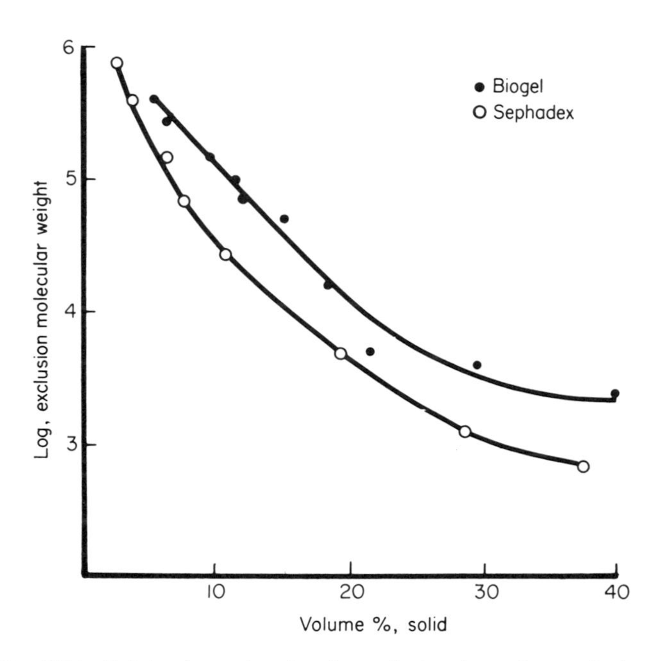

FIG. IV.1. Gel density and molecular exclusion (manufacturer's data).

porosity could be utilized by any one solute. However, such gels have some usefulness in molecular weight determinations of large polymers.

In most cases, gels appear to have a dispersion of pore sizes below a maximum value. In the process of gel preparation any factor which causes either the formation of nonuniform particles or particles of nonuniform composition would cause increased dispersion. Data of the type required to compare different lots of the same gel type are generally not available. In the utilization of gels for either molecular weight determination or fractionation each different lot number should be individually calibrated. Mixing gel lots of the same polymer type may result in increased dispersion.

2. *Functionality*

In principle, completely neutral gels containing no acidic or basic groups are most ideal for gel-permeation chromatography. However, commercially available gels contain a low concentration of ionizable groups. The cross-linked dextrans contain a few cation exchange sites (presumably –COOH) (*21*).

Fresh acrylamide gels appear to have few exchange sites. However, aging is apt to increase the exchange capacity of both types of gel.

The presence of exchange groups is deleterious to most separations.

Although completely neutral gels would not be expected to change volume in the presence of electrolytes, volume changes (due to change of osmotic pressure) do occur in the presence of electrolytes if ionized groups are present. If a volume change travels through the column the resulting band spreading can influence the separation. In addition, ionized groups can exert either exclusion or binding effects. Both effects have been reported for Sephadex gels (*22*).

3. *Solvent Binding*

Most gels appear to bind or complex a certain amount of solvent in such a way that it is not available to even the smallest electrolytes. Usually, in water the K_d obtained with NaCl or KCl is about 0.9, indicating that not all of the internal water is available to the electrolyte.

This small amount of unavailable internal fluid is not particularly important if the gel is being used for high molecular weight compounds. However, with the dense gels useful for fractionation of low molecular weight compounds this loss of available fluid may be an appreciable percentage of the total internal fluid. Since the volume ratio V_0/V_i is important in actual operation, control of column packing becomes more important when operating with dense gels.

III. COLUMN PERFORMANCE

Gel filtration, the group separation of molecules with large differences in size, and gel-permeation chromatography, the fractionation of molecules of similar size, are fundamentally quite different. Since column operating parameters have different optima, it is desirable to consider the two cases separately.

In gel filtration the excluded substance, $K_d = 0$, is separated from low molecular weight material which completely permeates the gel. Some internal solvent is usually unavailable and the K_d of low molecular weight materials is about 0.9. Since the excluded material does not permeate the stationary phase a small band spread and large charge volume is possible and low dilution is obtainable.

In gel-permeation chromatography the materials being separated permeate the gel, and diffusion coefficients within the gel contribute substantially to band spread. The charge volume is small and large dilution unavoidable.

The equation proposed by Giddings and Mallik (*23*) is useful in considering the problems of gel filtration and gel permeation. The general chromatography equation is considered in terms of factors contributing to the height of a theoretical plate. The column operation itself is considered ideal, i.e., mechanical effects, band spreading due to convection, sample and effluent distribution, and collection lines are not considered. Under these conditions the following expression for H (the height of a theoretical plate) has been proposed:

$$H = \frac{B}{v} + Cv + \frac{1}{1/A + 1/Fv} \tag{IV.3}$$

A, B, C, and F are constants and v is the linear velocity of the fluid passing through the column.

The first term, B, accounts for longitudinal diffusion. The complete term has been given as

$$B = 2D_{\mathrm{m}}\left[\gamma + \gamma_{\mathrm{s}}\frac{(1 - R)}{R}\right] \tag{IV.4}$$

in which D_{m} is the diffusion coefficient in the mobile phase; γ is an obstruction factor in the interstitial space and is assigned a value of 0.7; γ_{s} is the ratio $D_{\mathrm{s}}/D_{\mathrm{m}}$ in which D_{s} is the diffusion coefficient in the solid phase. This has a value in the range of 0.1–0.7. R is the retention or rate of movement of the solute zone relative to the mobile phase.

The second term, C, accounts for nonequilibrium effects in the stationary phase and for spherical particles has been expressed as

$$C = \frac{R(1 - R)\, d_{\mathrm{p}}^{2}}{30\, D_{\mathrm{m}}\, \gamma_{\mathrm{s}}} \tag{IV.5}$$

in which d_{p} is the average particle diameter.

The third term is a summation of flow pattern and nonequilibrium effects in the mobile phase. It is given as

$$\sum = \left[\frac{1}{2\, d_{\mathrm{p}}\lambda} + \frac{D_{\mathrm{m}}}{\omega\, d_{\mathrm{p}}^{2} v}\right]^{-1} \tag{IV.6}$$

In this expression λ and ω are structural factors which account for nonuniform packing and particle shape. For the present discussion, a value of unity is assigned to these constants.

Thus, for any particular chromatographic system, the terms A, B, C, and F are constant and H varies with velocity.

A. Application to Excluded Chromatographic System Materials

Elimination of diffusion effects within the stationary phase (C) and substitution in the individual terms leads to the following simplified expression:

$$H = \frac{2D_{\mathrm{m}}}{v} + \left[\frac{1}{2\, d_{\mathrm{p}}} + \frac{D_{\mathrm{m}}}{d_{\mathrm{p}}^{2} v}\right]^{-1} \tag{IV.7}$$

R is assumed to be 1. For excluded molecules D_m is on the order of 10^{-6} cm^2/sec (see Table IV.4) and the first term can be neglected for flow rates as low as 10^{-3} cm/sec. Since small molecules have tenfold greater diffusion coefficients, flow rates must be kept greater than 10^{-3} cm/sec to minimize band spread of these substances. The net value for H approximates

$$H = 2\, d_p \tag{IV.8}$$

For 100–200 mesh resin, H is then about 0.2 mm. Since in this simplified expression both the packing factor and the structural factor, have been assumed to be uniform, highly disperse, nonuniform particles and poor packing will cause increased H. For an excluded material $p = 1$, and σ in the same units as column length can be reduced to

$$\sigma = (L2\, d_p)^{\frac{1}{2}} \tag{IV.9}$$

in which L is column length.

The total band spread for an excluded substance is then the fraction of the mobile phase displaced by the sample plus (for 99.7% removal) 6σ. This indicates that arbitrary statements as to the volume of charge are useless in

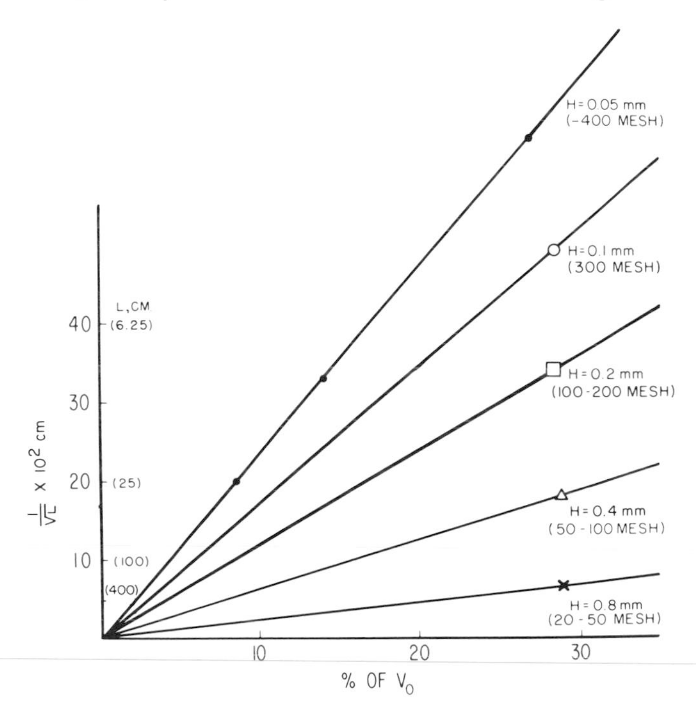

FIG. IV.2. Volume occupied by excluded materials; theoretical 6σ band spread.

the absence of additional information, especially column length and gel particle size. The percentage of mobile phase occupied by a band spread of 6σ for various particle sizes and column heights is plotted in Fig. IV.2. By assuming that the effective particle size is the average value, the sizes depicted correspond approximately to 50–100, 100–200, 200–400, and sub-400-mesh resin and no factor for nonuniformity has been included. A value of 0.4 of the bed volume was assigned to V_0.

The observed band spread includes mechanical factors of distribution and collection and mixing in the collecting equipment after the column. These contributions to band spread are rarely adequately described in experimental accounts. However, the band spread values obtained by Flodin (*24*) appear to correspond to those calculated by Eq. (IV.9) within a factor of 2–4. The recent work of Heitz and Coupek (*52*) offers a further, more elegant confirmation of the predictions of the theory.

If a factor of 3 is allowed as reasonable and usually experimentally obtainable with nonjacketed columns and dense feed solutions the dilution of a completely excluded substance can be estimated from Fig. IV.2. Considering a 100-cm column packed with 100–200 mesh gel, the 99.7% band spread is about 25% of V_0. The total volume of the excluded fraction is then $\frac{1}{4} V_0 + V_s$. If V_s is equal to $\frac{1}{4} V_0$ a twofold dilution would be obtained and if equal to $\frac{1}{8} V_0$ or $\frac{1}{2} V_0$, the dilutions would be 3-, and 1.3-fold, respectively.

B. Application to Completely Permeable Materials

Clearly for complete separation of high and low molecular weight materials the maximum charge volume depends on the band spread of the imbibed substance. Referring to Eqs. (IV.4) and (IV.5), it is apparent that the B term cannot be neglected and that contributions from the C term which includes diffusion in the stationary phase may be significant.

Insufficient data are available to assign an accurate value to γ_s. The obstructive properties of the gel may differ for different molecules (especially if electrostatic forces are present). In addition, the imbibed solvent may be different from the mobile solvent in internal organization so that different inherent diffusion rates would be obtained internally. Values of D_s/D_m as high as 0.7 and as low as 0.01 have been reported. Assuming that the range 0.1–0.7 applies to most cases and that $R = 0.5$ for permeable materials the B term reduces to

$$B = 2D_m\,[0.7 + (0.1 \text{ to } 0.7)] = 1.6 \text{ to } 2.8D_m \qquad \text{(IV.10)}$$

At an average value of 2.2 D_m and for a low molecular weight substance with a diffusion coefficient of 10^{-5} cm^2/sec, the contribution of this term to $H\,(B/v)$ is about 0.2 mm at $v = 10^{-3}$ cm/sec.

Similarly substitution into the C term gives the following value:

$$C = 1.2 \text{ to } 8.3 \times 10^{-3}\, d_p^2 v \tag{IV.11}$$

For resin of 100-μ particle size and a flow rate of 10^{-3} cm/sec C has a value of 0.001–0.008 mm.

Optimizing the B and C terms as a function of velocity it is found that

$$v = \frac{4.4 \text{ to } 15.2 \times 10^{-5}}{d_p} \text{ cm/sec} \tag{IV. 12}$$

The low value applies if γ_s is 0.1 and the high value if γ_s is 0.7.

The final term is also velocity-dependent and increases toward a value of $2d_p$ as velocity increases. The calculated plate height values representing the sum of all terms are plotted in Fig. IV.3. Two values of γ_s and four different

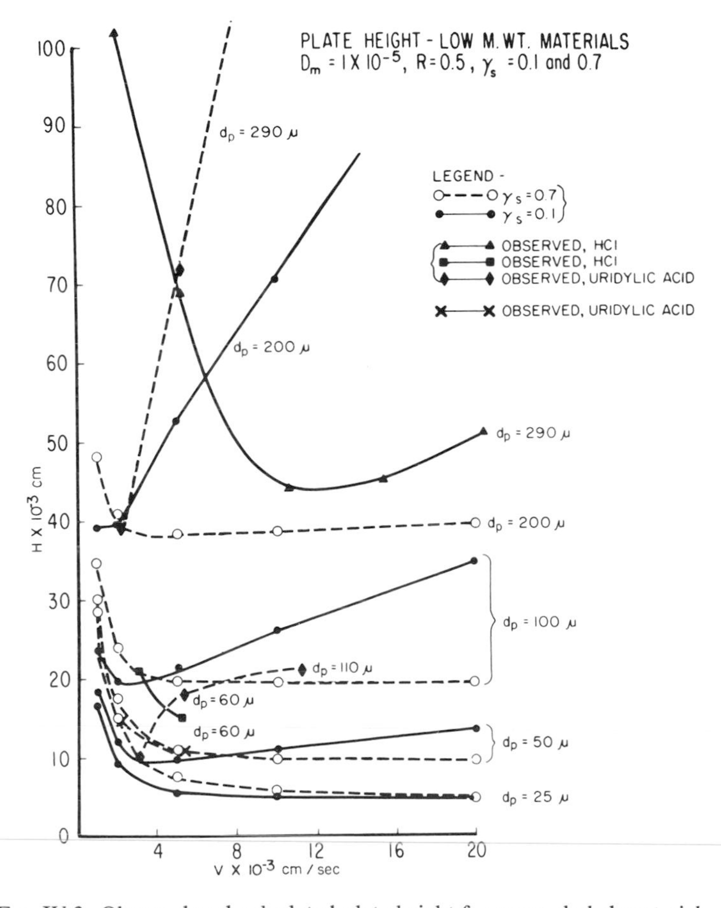

FIG. IV.3. Observed and calculated plate height for nonexcluded materials.

particle sizes are included. It can be seen that at or near the optimum velocity an average value for H is about $2d_p$. Although the value assumed for γ_s does have a pronounced effect on the net plate height, especially at lower velocity, the net effect is less significant for higher values of γ_s.

As a general practice it is desirable to use flow rates indicated by lower values of γ_s. This leads to the situation that finely divided resin, with which high flow rates are difficult to obtain, loses efficiency, whereas coarser resin attains maximum efficiency at relatively low flow rates.

Since resolving power is a function of $(1/H)^{\frac{1}{2}}$, columns having the same resolving power are shorter when packed with fine mesh resin and attain maximum efficiency at higher flow rates than columns packed with coarser resin. For example, a column prepared with 50-μ resin would be one-half as long and could be operated at 10 times the linear flow rate of a column prepared with 200-μ resin. Therefore, a 20-fold reduction in the amount of resin required for the same production rate appears possible. Uniform resins of small particle size are difficult to obtain and the mechanical problem of obtaining the desired flow rate is serious since even the most dense resins appear to have some compressibility under pressure.

C. Experimental Observations

Experimental data which can be compared with the above calculations are scanty. Since band spreading can take place in applying the sample and in the collecting and measuring equipment, experimentally observed values will always be greater than those calculated. A systematic study of the effect of particle size and flow rate on the height of a theoretical plate has been reported by Flodin (*24*). Different values were obtained for hydrochloric acid and uridylic acid. The data are plotted in Fig. IV.3. The coarse resin had an average particle diameter of about 290 μ, and the minimum H observed, about 450 μ, appears to be only 1.6 d_p. However, the spread in particle size, from about 220 to 370 μ may indicate that both λ and ω do not have values of unity. In addition, the use of an average for the particle diameter may not represent the actual material in the column. Although the minimum value for H agrees well with that predicted by the theory, the velocity maximum appears seriously off. The general shape of the H vs. velocity curve for hydrochloric acid would suggest that γ_s is more nearly 0.7 than 0.1. However, the same columns were used for uridylic acid, a substance which also permeates the gel. When compared using the coarse resin this compound appears to have a γ_s of even less than 0.1. The resin with the two remaining particles sizes investigated gives data in reasonably close agreement with that predicted for $\gamma_s = 0.1$.

Thus, in a general way the prediction of the theory, that effective plate height varies with velocity and that optimum values would be in the range of $2d_p$, appears correct. Unfortunately the uncertainty of the γ_s values indicates the necessity to determine optimum flow rates for the particular system being studied. The experimental data of Flodin indicate a marked dependence of minimum plate height on velocity for any specific column.

The use of resin of uniform particle size in the smallest size available seems indicated.

D. Sample Volume and Flow Rate for Group Separation

Assuming that flow rates can be kept within the range of $\frac{1}{2}$–2 times the optimum the band spread for substances with K_d values of about 0.9 and diffusion coefficients of 10^{-5} cm^2/sec is then approximated by

$$\sigma_s = 2(L2\, d_p)^{\frac{1}{2}} \tag{IV.13}$$

in which σ_s is expressed in the same units as column length. Since σ for both the excluded material and the low molecular weight material is known, the maximum sample volume may now be determined in terms of required resolution. A hypothetical elution diagram is depicted in Fig. IV.4.

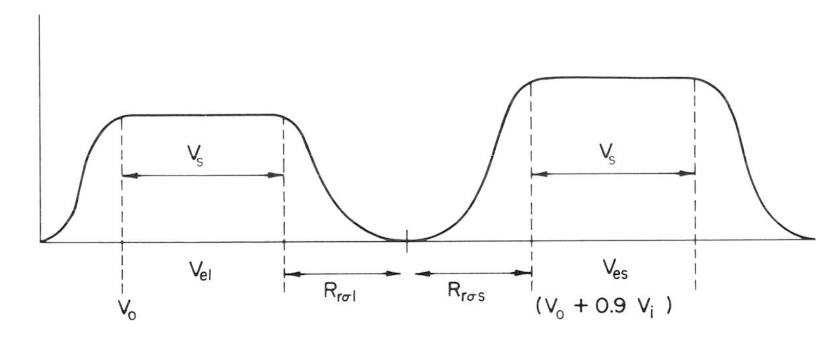

FIG. IV.4. Ideal gel filtration elution curves for completely excluded and completely permeable materials.

From the figure it can be seen that

$$V_{es} - V_{el} = R_r(\sigma_1 + \sigma_s) + V_s$$

and

$$V_{es} = 0.9V_i + V_0 + \tfrac{1}{2}V_s$$

$$V_{el} = V_0 + \tfrac{1}{2}V_s$$

Hence,

$$V_s = 0.9V_i - R_r(\sigma_1 + \sigma_s) \tag{IV.14}$$

in which R_r, relative resolution, is the distance in variances of the Gaussian

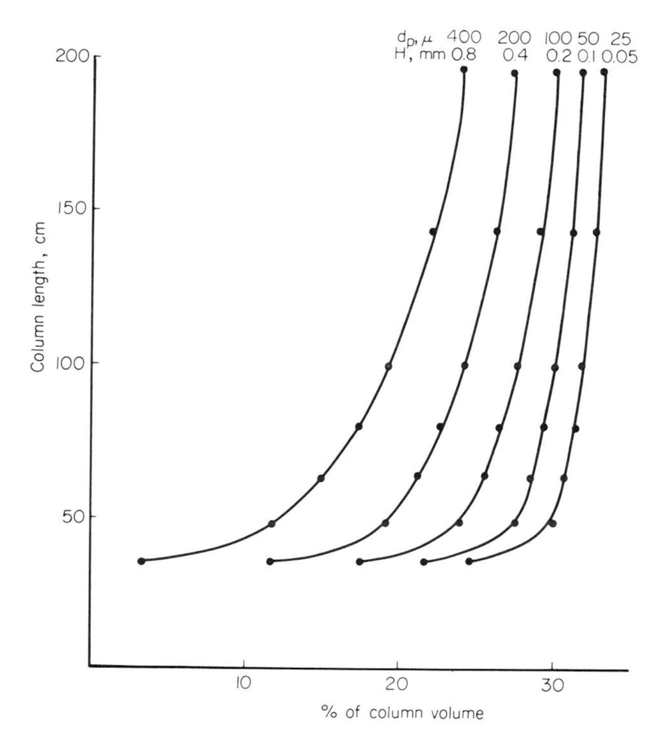

Fig. IV.5. Charge volume as influenced by column length and particle diameter; calculated as a function of column length and height of a theoretical plate.

curve from the imaginary sample edge. If $R_r = 2$, 97.5% of each component is separated, and if $R_r = 3$, 99.9% separation is obtained. For resins usually useful in group separation, V_i is about 0.4 of a column volume. Expressing the sample volume in terms of column length, L_{vs}, occupied by the sample the result is

$$L_{vs} = 0.36L - 3R_r(L2\,d_p)^{\frac{1}{2}} \tag{IV.15}$$

A plot of this relationship in terms of percentage of the total column volumn which can be occupied by the sample is depicted in Fig. IV.5. For this figure a value of 2 has been assigned to R_r. The figure can be used as a guide in determining the approximate column dimensions or sample volume to be used. For any column, experimentally determined values of H (instead of $2d_p$) and V_0 can be used for a more accurate determination of V_s.

E. Gel-Permeation Chromatography

Gel-permeation chromatography applies to separation of materials which partially permeate the column matrix, $0 < K_d < 1$. The entire separation must therefore take place within an eluate volume less than the internal volume, V_i, of the column. This is within the limits of V_{el} and V_{es} of Fig. IV.4. Giddings (*25*) has suggested that complete separation of more than five components, equally spaced, cannot be expected in the usual gel-permeation chromatography. Since most mixtures do not contain equally spaced components and since the procedure is frequently applied to crude mixtures, gel-permeation chromatography cannot be considered as a substitute for methods based on more specific properties of the substances being separated.

1. Plate Height

The previous discussion of the effects of diffusion rates and flow rates on plate height indicate that optimum flow rates are dependent on the diffusion coefficient of the materials being separated. The diffusion coefficients for a number of compounds are listed in Table IV.4 (*26*).

TABLE IV.4

DIFFUSION COEFFICIENTS AT 25°C

Substance	Molecular weight	D (cm²/sec × 10⁶)
Water	18	23.0
Urea	60	13.0
Glucose	180	6.8
Sucrose	342	5.0
Raffinose	504	4.3
Gramicidin S	1,140	2.6
Tyrocidin	2,473	2.1
Salmine	6,000	1.8
Myoglobin	17,200	1.1
Egg albumin	42,000	0.78
Hemoglobin	67,000	0.70
Edestin	294,000	0.43

For substances which permeate the gel and have diffusion coefficients in the range of $0.4\text{–}4 \times 10^{-6}$, the calculated height of a theoretical plate is dependent on the retention (variable between 0.5–0.8), the diffusion coefficient in the gel particle, and the particle diameter–flow rate ratio. The interplay of these can be resolved mathematically. However, it is probably more instructive to examine the trends in plate height introduced by each variable.

Within the range of $R = 0.6\text{–}0.8$ and $\gamma_s = 0.1\text{–}0.7$ with diffusion coefficients of $0.4\text{–}4 \times 10^{-6}$, the B term attains a value of 0.1 mm only at flow rates of less than 10^{-3} cm/sec. The corresponding development time is greater than 24 hr for a 100-cm column.

Examination of the components of the C term reveal the importance of diffusion in the gel particle, which is unknown. However, by assuming $\gamma_s = 0.1$, $D_m = 1 \times 10^{-6}$, and $R = 0.65$, plate heights of greater than 0.1 mm are obtained at flow rates faster than 2×10^{-4} cm/sec for resin of 100-μ size (100–200 mesh). Since this corresponds to a development time of about 6 days for a 100-cm column, this term seriously detracts from column efficiency. The plate height increases linearly with flow rate and is about 0.5 mm at a flow rate of 1×10^{-3} cm/sec.

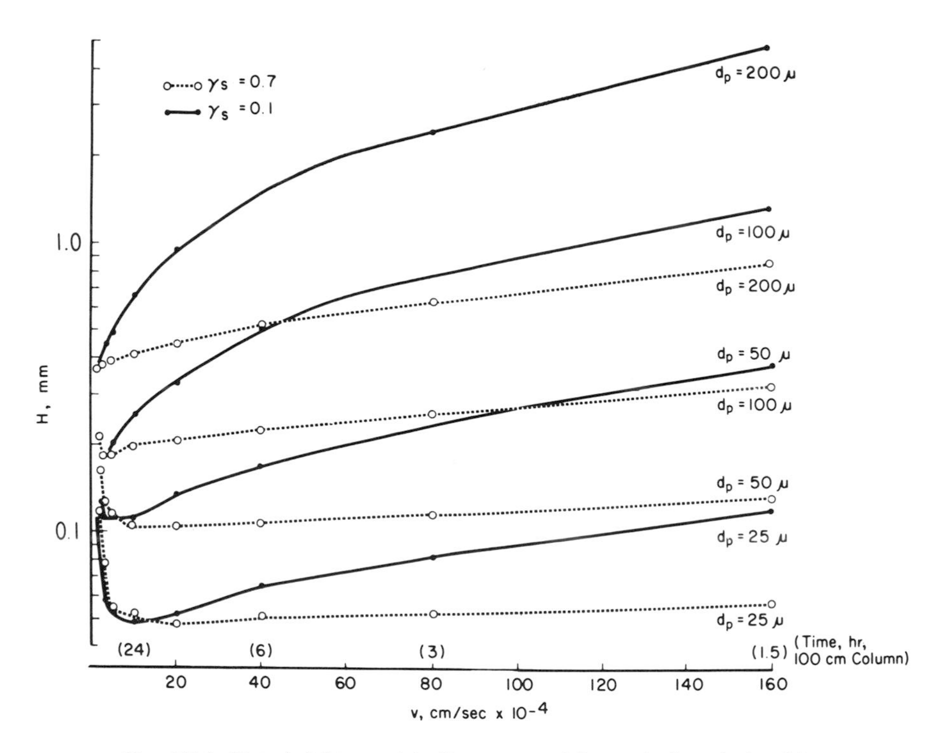

FIG. IV.6. Plate height, particle diameter, and flow velocity relationship.

The final term is dependent on particle size and varies with the flow rate. At a diffusion coefficient of 1×10^{-6}, the contribution to H is $0.7d_p$–$2d_p$ at the flow rates mentioned above.

These considerations are strictly theoretical and are based on geometric constants which, for the most part, have not been adequately studied in liquid systems. Performance data with gel-permeation chromatography columns are scanty but do not generally contradict the predictions of the theory which are summarized in Fig. IV.6. In this figure plate height is plotted as a function of flow rate for materials having a diffusion coefficient of 1×10^{-6} cm^2/sec. Two values for γ_s, 0.1 and 0.7, are included.

Examination of the figure illustrates that for a 100-cm column, flow rates corresponding to a development time of 6–24 hr should not drastically affect plate height except for relatively coarse resin. An approximate minimum for most cases is about $2d_p$.

2. *Resolution*

The resolution power of the method can be determined in terms of column length, particle diameter (or plate height), and K_d. Expressing σ in terms of particle diameter and column length,

$$\sigma_1 = \frac{1}{p_1}(L2\,d_p)^{\frac{1}{2}} \qquad \text{and} \qquad \sigma_2 = \frac{1}{p_2}(L2\,d_p)^{\frac{1}{2}}$$

since

$$p = \frac{V_0}{V_e} = \frac{V_0}{V_0 + K_d V_i}$$

let

$$\bar{p} = \frac{p_1 + p_2}{2}$$

Then

$$V_{e,1} - V_{e,2} = R_r(\sigma_1 + \sigma_2) = R_r \frac{2}{\bar{p}}(L2\,d_p)^{\frac{1}{2}} \tag{IV.16}$$

and

$$V_i(\Delta K_d) = R_r \frac{2}{\bar{p}}(L2\,d_p)^{\frac{1}{2}}$$

in which V_i is expressed in units of column length. V_i is normally about 0.6 of the total column volume. Therefore

$$\Delta K_d = \frac{3.3R_r}{\bar{p}}(L2\,d_p)^{\frac{1}{2}} \tag{IV.17}$$

By assuming that $\bar{p} = 0.7$, the approximate midpoint of a gel-permeation separation, ΔK_d as a function of percentage of column length is plotted in Fig. IV.7. The figure has been prepared as a function of plate height rather than particle diameter. Although information reported by Flodin (*24*) indicated plate heights as low as $2d_p$, other investigators have reported plate heights as high as $15d_p$ (*4*, *22*). The actual plate height attained in any chromatographic

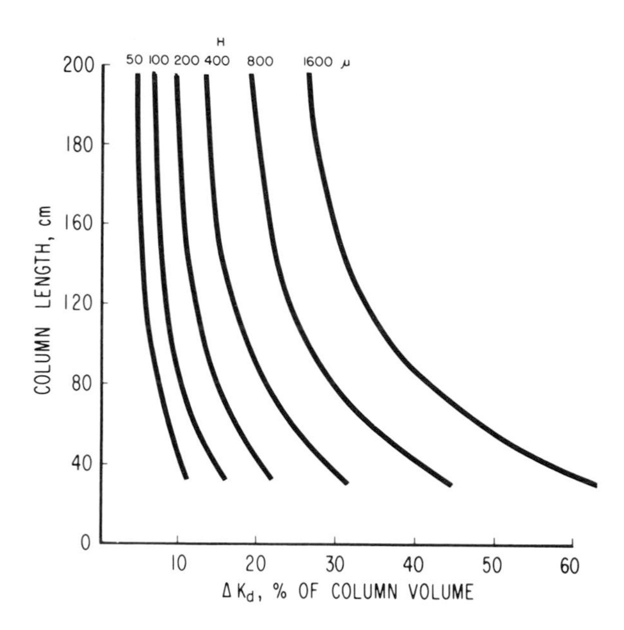

FIG. IV.7. Resolution in terms of column volume; ΔK_d at different plate height, $R_r = 2$.

system can be determined experimentally under the conditions to be used in the column. The width of the band containing 95% of a pure substance and the elution volume are related to the number of plates in the column by the relation

$$r = 16\left(\frac{V_e}{V_b}\right)^2 \tag{IV.18}$$

and

$$H = \frac{L}{r} = \frac{L}{16}\left(\frac{V_b}{V_e}\right)^2 \tag{IV.19}$$

Use of this figure together with molecular weight vs. K_d data for proteins or peptides indicates that the resolution of gel-permeation chromatography is generally poor. For example, a 100-cm column operated at a plate height of 0.2 mm gives an indicated ΔK_d equivalent to 13.3% of column length. Andrews (*3*) has shown that with Sephadex G-200 a K_d of about 0.4 is obtained with compounds having a molecular weight of 90,000. In terms of column volume, assuming $V_0 = 35\%$ and $V_i = 60\%$ of total volume, this material would emerge at $35 + (60 \times 0.4) = 59\%$ of the column volume. The nearest completely separated neighbor would be $35 + (24 \pm 13) = 48\%$ or 72% of the column volume. The corresponding K_d values, 0.2 or 0.6, give indicated molecular weights of 28,000 or 150,000, respectively. Thus it is clear that gel-permeation chromatography alone cannot be expected to yield pure material, no matter how carefully conducted.

The maximum resolution obtainable with a particular system is dependent on the resin utilized. In general, the most dense gel that can be used for the desired substance should be used since the same ΔK_d usually corresponds to a smaller difference in molecular size. With highly swollen gels the elasticity of the particles limits the flow rates obtainable.

In spite of the poor resolution usually obtained, gel-permeation chromatography constitutes a valuable addition to the tools available for isolation of natural products, especially when used in conjunction with other fractionation procedures. Even though none of the systems may provide high resolution of the compounds being isolated, a combination of procedures which rely on different properties of the compounds being fractionated may yield highly purified products. Gel-permeation chromatography can also be used for fractionation of substances whose K_d does not depend on size alone.

F. Factors Contributing to Column Performance

For both gel filtration and gel-permeation chromatography the calculations discussed above are useful only as guide lines indicative of good or poor column performance. Many factors may detract from this optimum situation. The pronounced variation in reported plate height has been mentioned and possible contributing factors discussed by Giddings (*25*). Since achievement of maximum efficiency can be of great importance, a number of factors which detract from optimum performance are discussed.

1. Flow Rate

The influence of flow rate on column efficiency has been mentioned above. Since chromatographic equipment usually differs in mechanical details such as column length, column feed arrangement, monitoring, and collection devices, the flow rate utilized is frequently a matter of convenience. The operating flow rate may be determined by such incidentals as the number of fractions which can be collected in an overnight run or the length of the recorder chart which would be produced. In some cases the flow rate is limited by the nature of the gel being used. Unstable compounds may require rapid operation.

If the gel step is critical in the fractionation procedure, a determination of the effect of flow rate on plate height, using readily detectable substances with K_d values similar to the compounds being fractionated, in the actual column used is recommended.

Control of the flow rate is especially important in scaling-up from small to larger columns.

2. Sample Size, Viscosity, and Density

The effect of sample size on the total band spread for a particular component has been mentioned. In the processing of crude materials, reduction of sample size by concentration can produce solutions having undesirable viscosity or density properties. Solutions of high molecular weight materials may be quite viscous. Development of columns charged with viscous solutions frequently produces band spreading due to "fingering," a phenomenon in which pockets of the viscous solution are by-passed by the less viscous developing solution. If this occurs good separation is virtually impossible. In addition, diffusion coefficients decrease as viscosity increases. In the kinetic system, the resulting increase in the nonequilibrium terms causes increased band spreading.

Density differences between the sample and the developing solution may cause both increased band spread and displaced elution peaks. If the developing solvent is either water or buffer at low concentration, solute bands more dense than the solvent are formed. This difference causes the movement of the solute bands to be more rapid than solvent movement and changes the K_d. If the column is not mounted exactly plumb, the band movement due to density has a lateral vector which further increases band spread.

The change in K_d may either detract from or enhance the actual separation, depending on whether the leading zone is less or more dense than slower moving zones.

Good performance of gel-filtration columns can be obtained with samples containing as high as 50% solids.

Samples containing high concentrations of excluded substances may cause shrinking of the gel beads. If a volume change is obtained, both band spreading and altered K_d of imbibed substances may reduce resolution.

3. Sample Application

The introduction of the sample is of greater importance in gel-permeation chromatography than in chromatographic methods in which the solute is adsorbed, since the latter process may produce a narrower band than existed in the sample. With gel-permeation chromatography this is not possible. It is therefore necessary to introduce or apply the sample evenly in as narrow and uniform a band as possible.

Applying a sample to the top of the column by means of a pipet held against one side of the column while withdrawing liquid from the bottom produces an uneven zone which persists through the entire column. If a sample layer is built up on top of the column the addition of developing solvent can cause dilution which then persists through the run.

Samples can be introduced as a band under the developing solution with the column exit off and then slowly drawn into the bed before starting development. Sintered discs placed on top of the resin bed may help obtain even sample distribution. If a liquid head is maintained above the gel, a plastic float may be used to disperse entering drops of solution.

Columns equipped with adjustable inlet lines which apply the sample directly to the resin bed or to a sintered disc on top of the bed are available. If the inlet stream spreads evenly across the gel bed this method of sample injection is more reproducible than many other commonly accepted methods.

4. *Solubility and Aggregation*

Separation of macromolecules may be impaired if the solute becomes insoluble due to separation of solubilizing electrolytes or pH changes during the separation. Although this can usually be overcome by the use of buffers of adequate ionic strength and of the proper pH for the desired substance, any precipitation causes increased band spreading. In some cases molecular aggregates form which are either pH-dependent or concentration-dependent. Such a change in the nature of the material being separated causes increased band spread and lack of resolution.

5. *Temperature Control*

Lack of temperature equilibration during a chromatographic separation causes increased band spreading. The predominant effect is due to distortion of the liquid flow pattern within the column. For this reason the use of jacketed columns and temperature equilibration of the developing solution is recommended.

6. *Bubble Formation*

Bubble formation due to release of gas within the column can cause serious channeling. Columns containing an air space above the developing solvent which is subject to operating pressure may release gas in the lower part of the column due to pressure changes and diminished solubility.

Growth of microorganisms within the column may release gas. This may occur even in the cold since gel columns frequently contain buffers and organic compounds susceptible to attack by microorganisms.

7. *Sample Collection*

Many columns are constructed with a dead space between the bed support and the stopcock or flow regulating device. If this space is filled with liquid much efficiency may be lost. In addition the lines leading to monitoring equipment, the flow cells in such equipment and/or lines to a fraction collector can cause band spread if the total hold-up volume is large compared to the volume of a theoretical plate.

This mixing in the auxiliary equipment may cause much more loss of resolution than is suspected.

IV. FACTORS INFLUENCING K_d

If the only fact which affected K_d were the molecular size and shape of the molecular species involved, the K_d would not be changed by any operational factor, and the relative K_d values of two compounds (β value or separation factor) would be independent of electrolyte background, electrostatic effects, or interaction with polymer backbone. Although this is true for many cases, it has also been well established that relative K_d values can be affected by properties of the system other than those attributable to a neutral gel. Some of the factors influencing K_d are discussed in the following sections.

A. Flow Rate

Separation by gel filtration is a kinetic process. In the above discussion concerning flow rates, it has been assumed that the time required to achieve nearly complete equilibration of any solute between the mobile phase and the internal liquid of the gel particles is somewhat less than the average contact time. If this is not the case and the flow rate is of such speed that the mobile phase always contains a much greater concentration of solute than the equilibration value, then the solute will begin to emerge from a column operated at high speed earlier than if the column is operated at a lower speed. Using a simplified model of a chromatography column based on diffusion, Vink (*27*) presented calculations which indicate that K_d is diminished as velocity increases and band spread increases simultaneously. Laurent and Laurent (*28*) constructed an analog computer in which capacitors represented the individual plates of a chromatographic column. By varying the rate of charge of the capacitors and the number of charges introduced, the effect of column flow rate and volume charged on the shape and position of the elution band was determined. The data obtained with this computer also indicate decreasing K_d and increasing band spread as flow rate increases.

Flodin (*24*) studied the effect of flow rate on K_d and band spread.

Using uridylic acid and Sephadex G-25, decreased K_d and increased band spread were obtained as the flow rate increased using a particular column and resin particle size. At the same flow rate, similar effects were obtained as the average resin particle size was increased.

Examination of the data of Whitaker (*4*) indicates the possibility that K_d values are influenced by flow rate.

B. Adsorption

The interaction of the resin phase with compounds containing aromatic nuclei has been well documented and occurs with all types of gels, albeit with differing intensity. The effect with Sephadex gels is illustrated by the data of Rutenberg and Craig (*29*) obtained with the tyrocidin antibiotic complex. In this study it was shown that increasing acetic acid concentration diminished the adsorption. The elution volume obtained with 50% acetic acid–water corresponded to that expected on the basis of molecular size and no separation of the complex was obtained. When the developing solvent was water alone, very strong retention of the tyrocidin complex was obtained and resolution was not possible. At intermediate acetic acid concentrations the antibiotic could be separated into its individual components. In this system the compounds containing the most tryptophan are the slowest moving.

A solvent mixture of phenol–acetic acid–water has been recommended to eliminate adsorptive binding of peptides to Sephadex G–25 (*30*).

In a carefully conducted study Marsden (*1*) showed that the retention of simple aliphatic primary alcohols and low molecular weight ketones on a dense dextran cross-linked with divinyl sulfone groups increased markedly as the polarity of the molecules decreased. The hydrocarbon chain may interact with the polymer backbone or the compounds may be more "soluble" in the internal liquid of the gel than in the mobile phase.

Simple amides, such as urea, thiourea, and formamide, are also strongly retarded by dextran gels.

C. Ion Exchange

Most gels contain ionizable groups. These may be attached by a covalent bond to the polymer chain or merely strongly adsorbed within the gel matrix. Covalently bound exchange groups are usually carboxylic and may result from traces of acrylic acid in acrylamide gels or oxidized terminal groups in dextran gels. Further hydrolysis or oxidation which may occur on use would increase the number of exchange groups.

Strongly bound substances having ionizable groups may be present in the gel. These may be introduced during manufacture or acquired in subsequent use. In a detailed study of the effects of the developing solvent on K_d, Eaker and Porath (*22*) reported significant changes after washing the gel (Sephadex G–10) with pyridine. The strongly bound pyridine was removed with acetic acid. After pyridine washing the K_d of basic amino acids decreased. It was postulated that the pyridine washing had removed sulfate ester detergents used in the gel manufacture. Thus, the nature and extent of ionizable groups present in the gel may reflect previous treatment of the gel.

The effect of ionized groups within the gel is generally to bind opposite types or exclude similar types. For this reason, the K_d of ionizable substances may vary with pH changes which influence either the ionization of the solute or the gel phase.

These effects are illustrated by a study of the gel-permeation chromatography of a series of peptides, amino acids, and dinitrophenylamino acids using a highly cross-linked dextran (*31*). Some of the compounds studied were: vasopressin, oxidized glutathione, polymixin, isoleucine, glycine, dinitrophenylisoleucine, dinitrophenylglycine and arginine.

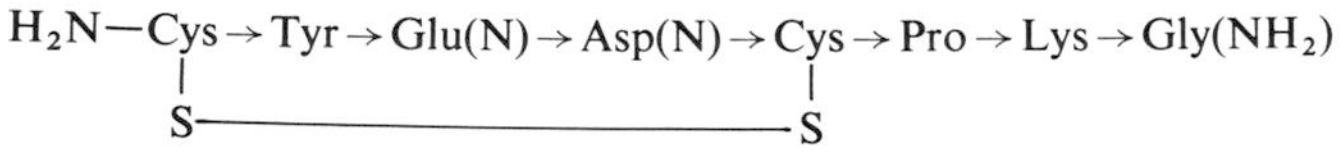

$$
\begin{array}{l}
H_2N\text{—}Cys \rightarrow Tyr \rightarrow Glu(N) \rightarrow Asp(N) \rightarrow Cys \rightarrow Pro \rightarrow Lys \rightarrow Gly(NH_2) \\
\quad\; | \qquad\qquad\qquad\qquad\qquad\qquad\qquad\qquad | \\
\quad\; S\text{————————————————}S
\end{array}
$$

Vasopressin

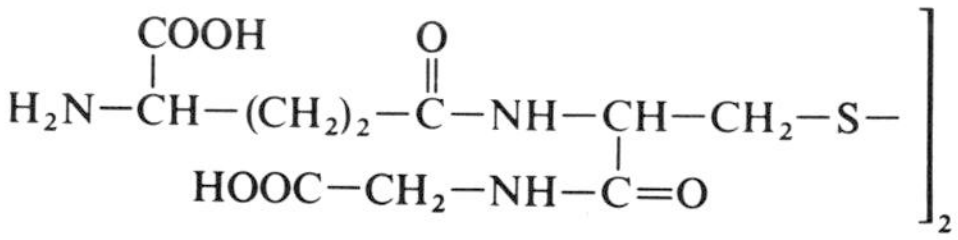

$$
\left[
\begin{array}{l}
\qquad\;\; COOH \qquad\qquad\quad O \\
\qquad\;\;\; | \qquad\qquad\qquad\quad\; \| \\
H_2N\text{—}CH\text{—}(CH_2)_2\text{—}C\text{—}NH\text{—}CH\text{—}CH_2\text{—}S\text{—} \\
\qquad\qquad\qquad\qquad\qquad\qquad\quad\; | \\
\qquad\qquad HOOC\text{—}CH_2\text{—}NH\text{—}C{=}O
\end{array}
\right]_2
$$

Oxidized glutathione

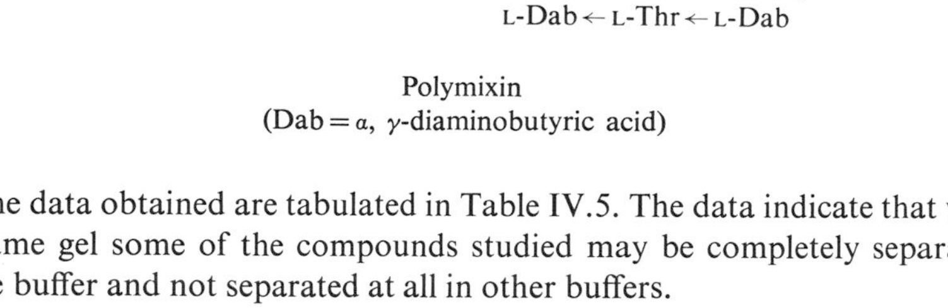

$$
\begin{array}{l}
\qquad\qquad\qquad\qquad\qquad\qquad\qquad \text{L-Dab} \rightarrow \text{D-Phe} \rightarrow \text{L-Leu} \\
\quad O \qquad\qquad\qquad\qquad\qquad\quad \nearrow \qquad\qquad\qquad\qquad \searrow \\
\quad \| \\
R\text{—}C \rightarrow \text{D—Dab} \rightarrow \text{L-Thr} \rightarrow \text{L-Dab} \qquad\qquad\qquad\qquad\qquad\qquad \text{L-Dab} \\
\qquad\qquad\qquad\qquad\qquad\qquad\quad \nwarrow \qquad\qquad\qquad\qquad \swarrow \\
\qquad\qquad\qquad\qquad\qquad\qquad\qquad \text{L-Dab} \leftarrow \text{L-Thr} \leftarrow \text{L-Dab}
\end{array}
$$

Polymixin
(Dab = α, γ-diaminobutyric acid)

The data obtained are tabulated in Table IV.5. The data indicate that with the same gel some of the compounds studied may be completely separated in one buffer and not separated at all in other buffers.

D. Chemical Reactivity

Although the substances used for gel filtration are relatively inert, functional groups are present and may react with certain types of molecule. Thus, in the presence of cyanogen halides, proteins are irreversibly linked to dextran gels (*32*). The amide groups of acrylamide polymers may combine with

TABLE IV.5

R_f OF VARIOUS SUBSTANCES WITH A DEXTRAN GEL

Substance	Buffers used[a]			
	1	2	3	4
Vasopressin	0.56	0.55	0.51	0.30
Polymixin	0.86	—	0.65	0.15
Oxidized glutathione	0.61	0.66	—	0.99
Isoleucine	0.52	0.51	0.50	0.50
Glycine	0.51	—	0.50	0.48
DNP-isoleucine	0.28	—	0.36	0.77
DNP-glycine	0.26	—	0.34	0.77
Arginine	0.52	—	0.41	0.30

[a] Buffers: 1, 1 *M* acetic acid; 2, 0.3 *M* pyridine + 0.1 *M* acetic acid 3, 1 *M* pyridine + 0.05 *M* acetic acid; 4, 1 *M* pyridine.

compounds containing a reactive carbonyl group and at high pH de-cross-linking and liberation of reactive methylol groups may occur.

E. Secondary Binding or Exclusion

The interaction involved is that which occurs between solutes in such a manner that the K_d of one, but probably not both, of the reacting components is changed markedly. This is dramatically illustrated by a recovery procedure for trace amounts of vitamin B_{12} present in seawater (*33*). In this procedure a protein from pig pyloric mucosa which binds vitamin B_{12} (m. wt., 1300) is added to the sample. The protein–vitamin B_{12} complex can then be separated from the seawater salt by gel filtration. The resulting salt-free solution can be concentrated and assayed. Gel filtration with the resin used does not separate vitamin B_{12} and salt.

The distribution of small electrolytes may be complex. When sodium chloride was added to 0.2 *M* acetic acid and chromatographed on Sephadex G-10, a band of higher pH, presumably sodium acetate, emerged prior to the band containing sodium chloride. The latter part of the salt band was at low pH and apparently contained hydrochloric acid (*22*).

Interactions of this sort may drastically enhance the separating power of the method and account for many successful separations of compounds of similar size.

Recently, Saunders and Pecsok (*34*) measured the K_d for a number of small electrolytes with BioGel P-2, an acrylamide gel. In most cases K_d values greater than unity were reported. In addition contributions of individual ions were shown to be additive. That is, the K_d of M^+X^- could be predicted from values for M and X obtained with other salts. For example, $BaCl_2$ and $KClO_4$ have K_d values of 1.64 and 1.77, respectively. An equimolar mixture gave two peaks and an excess of $BaCl_2$ gave three peaks. These findings are consistent with the presence of both cationic and anionic charge groups within the gel.

V. APPLICATIONS OF GEL FILTRATION

Gel-filtration separations are of a group type and generally use relatively dense gels. The procedure is frequently substituted for dialysis but can be applied to substances with molecular weights as low as 300. The operation is frequently more rapid, more complete, and more readily adapted to large-scale production than dialysis. Less dilution of the sample is frequently possible.

A. Desalting or Buffer Change

Many examples of desalting or buffer change by gel filtration are contained in the literature. It may be very useful in the work-up of ion exchange chromatography cuts obtained with high concentrations of salt. The procedure used for the separation of salt from chromatographic fractions of heparin is a typical example (*35*). In this case Sephadex G-25 was used. A column, 2 × 14 cm, was charged with 1 ml of concentrated column cut and developed at a flow rate of 3 ml/min (1.6×10^{-2} cm/sec). An entire separation required about 15 min. Complete separation of 2.5 mg of heparin and 110 mg of salt was reported and the dilution of the heparin solution was about sixfold.

For this type of problem, optimum column efficiency is relatively unimportant in comparison with the ease and speed of the separation.

Viscous solutions which do not flow readily through deep beds have been desalted using a basket centrifuge (*36*). In this procedure a layer of gel (Sephadex G-25) is built up inside the centrifuge basket. The void fluid is centrifuged out at relatively high speed. The centrifuge is slowed and a charge of about 15% of the bed volume is added. The centrifuge speed should be regulated so that at least 10 min is required for the charge to travel through the bed. The centrifuge speed is increased to remove the desalted solution from the bed. The salt is then rinsed out at slow speed. Although this procedure may be applied at times when other procedures cannot be operated mechanically, losses may be substantial, depending on the viscosity of the starting solution, the particle size of the gel, and the liquid hold-up in the centrifuge.

Desalting procedures with columns can be readily automated using either timed flow or conductivity sensors. The use of the latter for the separation of Ficoll, a sucrose polymer, and salt has been reported (*36*).

B. Separation of Reactants

The use of gel filtration to separate low molecular weight reactants or reaction products from reactions with macromolecules is illustrated in the preparation of carboxymethyl peptides or proteins. The reaction is carried out in 5–8 *M* urea with up to 1000 moles excess of mercaptoethanol and iodoacetic acid. At the completion of the reaction the urea and low molecular weight reactants are separated from the high molecular weight material by gel filtration. For this purpose Crestfield *et al.* (*37*) reported the use of Sephadex G-75 in 50% acetic acid.

Gel filtration is especially effective in cases in which traces of unreacted starting material remaining in the product seriously detract from the success of the method, and must be completely removed. Examples are fluorescent labeling of antibodies (*38*) and tracer labeling of proteins (*39*).

C. Determination of Complex Formation or Binding Ability

Gel filtration can be applied in the determination of complex formation or binding between permeable and impermeable substances. In the simplest application a mixture of two substances is charged to a column and the column effluent monitored for the appearance of the low molecular weight substance which occurs sooner if a complex is formed with the larger molecule.

Acred *et al.* (*40*) investigated the binding of three penicillins to serum proteins using both dialysis and gel filtration. Although the gel filtration data were not quantitative, the relative binding of the three compounds was the same by the two methods.

A novel procedure was used to determine binding between a low molecular compound, which does not react under the chromatographic conditions employed, and an enzyme (*41*). The column was equilibrated with the low molecular weight substance and a small amount of enzyme charged to the column dissolved in the same solution. Concentration profiles of the low molecular weight substance indicate an increase at the elution volume of the enzyme and a corresponding decrease at the elution volume of the bound material. The total amount of bound material can be estimated from the areas of both regions. In this case cytidylic acid was used as the low molecular substance and ribonuclease, but not inactive forms of ribonuclease, was shown to bind the compound.

Surfactants which form micelles at a critical concentration have been studied (*42*). At concentrations higher than the critical micelle concentration the micellar surfactant (sodium decylsulfate) is excluded from the gel. Added materials were shown to partition between the micelle and aqueous solution. Thus the elution volume of added *p*-nitrophenol was observed to be dependent on the concentration of sodium decylsulfate due to its preferential solubility in the micellar phase. Glycine amide, which does not enter the micelles readily, is not influenced by the presence of even high concentrations of sodium decylsulfate.

VI. APPLICATIONS OF GEL-PERMEATION CHROMATOGRAPHY

The process is used extensively in purification of natural products, especially macromolecules which may be unstable. The procedure cannot be considered a high resolution separation method for the reasons outlined previously. As a separation method, capacity is usually rather low but yields are generally high.

Many valuable biochemical separations which are difficult to achieve by other methods can be performed by gel-permeation chromatography. For example, cell extracts contain both ribonucleotide polymerase and ribonuclease enzymes. Since the nuclease activity predominates the polymerase cannot be readily detected until the two enzymes have been separated. Gel-permeation chromatography has been used in the separation of these enzymes (*46*).

In conjunction with other separation techniques the procedure has been used with peptide hormones, e.g., thyrocalcitonin. In this case the retention of the hormone is influenced by factors other than molecular size.

The use of mixed solvents has given separations which appear to depend both on size and polarity of the substances being separated. If the gel phase is equilibrated with a mixed solvent, the imbibed solvent may have a different composition than the mobile solvent. Under these circumstances partition between the phases may contribute to the separation.

A. Molecular Weight Determination

Gel-permeation chromatography has found rather extensive use in the estimation of molecular weights of macromolecules. The method was originally applied to compounds of biological interest using aqueous buffer solutions. More recently a completely automated system using solvents and hydrophobic gels has been applied to the estimation of synthetic polymer weights.

The procedure is fundamentally simple and requires only an accurate determination of the elution volume from a column chromatographic separation or relative mobility on thin-layer chromatogram. Calibration usually consists of determination of the elution volume or mobility of known compounds in the same system or simultaneously with the unknown. Although molecular size usually is the most important single property influencing elution volume, the shape of the molecule as well as any tendency to dimerization or aggregation can be of great importance. The procedure is best applied to spherical substances. Andrews (*3*) has pointed out that glycoproteins (such as γ-globulins) usually appear to have a larger molecular size than globular proteins. Actual molecular weights for many proteins must be considered still unknown in that reference values are of necessity based on some other physical measurement such as ultracentrifugation, etc.

1. Column Procedures

For proteins of molecular weight greater than 25,000, Andrews used Sephadex G-200 (200–300 mesh) equilibrated for at least 5 days with 0.05 *M* tris hydrochloride buffer, pH 7.5, and 0.1–0.15 *M* KCl. A column, 2.5 × 50 cm, was carefully packed at 2°–5°C. The feed solution was usually 2 ml containing about 1 mg of protein. At a flow rate of about 15–18 ml/hr (0.4–0.6 $\times 10^{-3}$ cm/sec) a single run was completed in less than 20 hr. Fractions of 3 ml were collected and the protein concentration determined by ultraviolet absorption or enzyme assay. The elution volume was estimated to the nearest milliliter by triangulation.

For nonaqueous systems a series of polystyrene gels of differing porosity which swell in organic solvents is available. Tetrahydrofuran, which swells the gel, has low viscosity and is a good solvent for many compounds of interest. It has been used extensively for the analysis of the molecular weight dispersion of polymers. The same system, which employs detection by refractive index, was used to determine the apparent molecular size of a number of small molecules. In this procedure (*19*) columns 8 mm × 240–480 cm, containing polystyrene beads (25–50 μ in size) were operated at a flow rate of 1 ml/min (4–6 $\times 10^{-2}$ cm/sec). The compounds studied include those containing one or more oxygen, nitrogen, halogen, phosphorous, or sulfur atoms. In general, the molecular size can be determined within 0.5 C atom if certain empirical corrections for aromatic or cyclic structures and solvent interaction, especially hydrogen bonding with HO-containing compounds, are applied.

Referring to Fig. IV.6 and extrapolating to the high flow rates used in this study, it can be seen that as diffusion in the particle becomes limiting plate heights increase rapidly. This effect was observed. A plate height of 3 mm was reported for dodecyl carbonate and only 0.5 mm for acetone. It would appear that considerably lower flow rates would be required to give good resolution of polymer mixtures of high molecular weight.

2. TLC Procedures

Several authors (*43*, *44*, *45*) have investigated the use of thin-layer chromatography (TLC) plates for molecular weight determination. Operational difficulties due to shrinkage of the gel on drying, fast flow rate, determination of the solvent front, and location of the protein have been overcome by using very fine resin, carefully storing and equilibrating prepared plates, development at an inclined angle, adding known markers, and by either staining paper "prints" of the developed chromatogram or use of the Rydon-Smith test on wet plates. Plates containing Sephadex G-25, G-50, G-100, and G-200 have been used and in each case satisfactory molecular weight extrapolations can be made.

B. Gel-Permeation Chromatography in the Purification of Peptide Hormones

Calcitonin is a peptide hormone produced by thyroid tissue. The hormone has a molecular weight of about 3600, and a net basic charge due to the presence of two arginine groups and only one carboxyl. Of the 32 amino acids 5 (3 phenylalanine, 1 tyrosine, and 1 tryptophan) are aromatic and might contribute to secondary binding.

Gel-permeation experiments have been reported with Sephadex G-100, G-75, G-50, Biogel P-6, and P-10 using starting material of differing potency. With Sephadex gels the hormone has a K_d of greater than 1 with most systems but retention is greater in neutral or alkaline systems than in acidic systems. With Biogel P-6 resins the retention is also influenced by the solvent used. An extrapolated molecular weight value about 3000 was

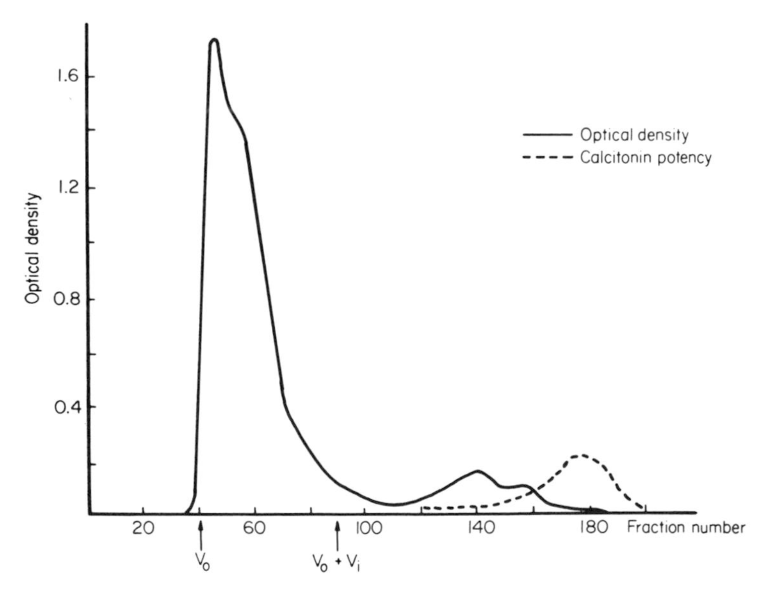

FIG. IV.8. Chromatography of calcitonin on Sephadex G-75; 0.2 *M* ammonium acetate buffer, pH 4.6.

obtained with 0.4 *M* acetic acid whereas with 0.1 *M* formic a value of 4300 was reported.

Using a crude preparation (about 0.1% pure) a column of Sephadex G-75 (2.5 × 130 cm) was charged with a solution of 500 mg in 5 ml. Development was 0.8 ml/min (7 × 10^{-3} cm/sec) and fractions of 6.4 ml were taken. The column was monitored by ultraviolet absorption and the fractions assayed for hormone content. The data obtained from such a run are summarized in Fig. IV.8 (*47*, *48*). Examination of the figure indicates that substantial purification has been achieved and that the product is probably not pure.

Using Biogel P-6 and 0.4 *M* acetic acid, a 2 × 225 cm column was charged with 1–2 ml of solution containing 10–50 mg of peptide. The elution volume of the hormone was compared with vitamin B_{12} (m. wt. 1300) and insulin (m. wt. 5700) and indicated a probable molecular weight of 3000 (*49*).

The use of gel-permeation chromatography has also been reported for the purification of beef parathyroid homone (*50*). Using Sephadex G-50, the elution volume was 16% of the total column volume with 0.2 *M* ammonium acetate and 40% of the total column volume using 0.2 *M* acetic acid.

The partial separation of the insulin A and B chains using Sephadex G-75 and 50% acetic acid has been reported (*51*).

C. Mixed Solvent Application

Using methylated Sephadex G-25, mixtures of choroform and methanol were investigated for the separation of cholesterol and a tetrahydroxycholane (*20*).

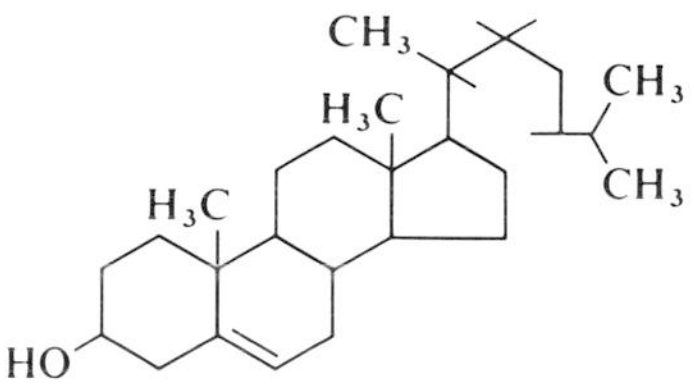

Cholesterol, m.wt. 387, $C_{27}H_{46}O$

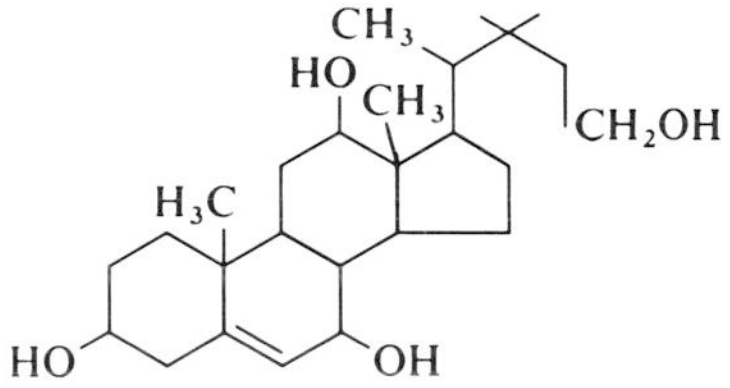

3α, 7α, 12α, 24α-Tetrahydroxycholane, m.wt. 392, $C_{24}H_{40}O_4$

The pattern of K_d values obtained at different percentages of methanol is plotted in Fig. IV.9. It is apparent that the factors influencing the K_d of the related molecules are complex but that by choosing the conditions carefully good separation can be obtained. The use of ethanol gives no separation of the two steroids. An apparent straight-line relationship between K_d and the solvent composition occurs for both compounds. The abrupt change in direction of these lines could be due to a relatively drastic change in the solvent most intimately associated with the polymer. A reversal in K_d trend in obtained. The greatest swelling was observed at 20% methanol concentration and may also indicate a type of phase change about this solvent composition.

This procedure, which has not been thoroughly studied, appears capable of achieving specific separations which might be difficult by other procedures.

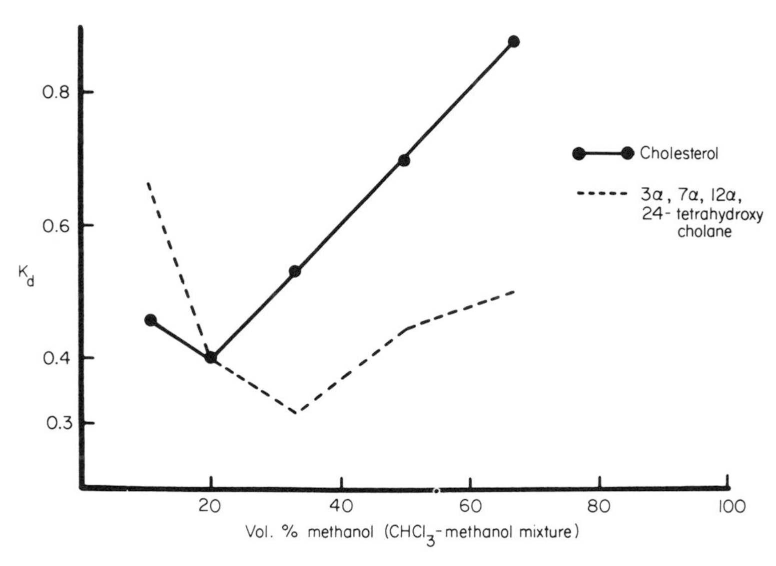

FIG. IV.9. Effect of mixed solvent on K_d. Separation of steroids in $CHCl_3$–methanol mixtures on methylated Sephadex G-25.

REFERENCES

1. Marsden, N. V. B., *Ann. N.Y. Acad. Sci.* **125**, 428–457 (1965).
2. Lathe, G. H., and Ruthven, C. R. J., *Biochem. J.* **62**, 665–674 (1956).
3. Andrews, P., *Biochem. J.* **96**, 595–606 (1965).
4. Whitaker, J. R., *Anal. Chem.* **35**, 1950–1953 (1965).
5. Ackers, G. K., *Biochemistry* **3**, 723–730 (1964).
6. Porath, J., *Pure Appl. Chem.* **6**, 223–244 (1963).
7. Laurent, T. C., and Killander, J., *J. Chromatog.* **14**, 317–330 (1964).
8. Squire, P. G., *Arch. Biochem. Biophys.* **107**, 471–478 (1964).
9. White, M. L., *J. Phys. Chem.* **64**, 1563–1564 (1960).
10. White, M. L., and Dorion, G. H., *J. Polymer Sci.* **55**, 731–740 (1961).
11. Haller, W., *Nature* **206**, 693–696 (1965).
12. Hjerten, S., and Mosbach, R., *Anal. Biochem.* **3**, 109–118 (1962).
13. Sun, K., and Sehon, A. H., *Can. J. Chem.* **43**, 969–976 (1965).
14. Flodin, P., "Dextran Gels and Their Applications in Gel Filtration." Pharmacia, Upsala, 1962.
15. Hjerten, S., *Biochim. Biophys. Acta* **79**, 393–398 (1964).

16. Polson, A., *Biochim. Biophys. Acta* **50**, 565–567 (1961).
17. Wheaton, R. M., and Bauman, W. C., *Ann. N.Y. Acad. Sci.* **57**, 159–176 (1953).
18. Brewer, P. I., *Nature* **190**, 625–662 (1961).
19. Hendrickson, J. G., and Moore, J. C., *J. Polymer Sci.* **4**, 167–188 (1966).
20. Nystron, E., and Sjovall, J., *Anal. Biochem.* **12**, 235–248 (1965).
21. Miranda, F., Rochat, H., and Lissitzky, S., *J. Chromatog.* **7**, 142–154 (1962).
22. Eaker, D., and Porath, J., *Separ. Sci.* **2**, 507–550 (1967).
23. Giddings, J. C., and Mallik, K. L., *Anal. Chem.* **38**, 997–1000 (1966).
24. Flodin, P., *J. Chromatog.* **5**, 103–115 (1961).
25. Giddings, J. C., *Anal. Chem.* **39**, 1027–1028 (1967).
26. Polson, A., and van der Reyden, D., *Biochim. Biophys. Acta* **5**, 358–360 (1950).
27. Vink, H. J., *J. Chromatog.* **15**, 488–494 (1964).
28. Laurent, T. C., and Laurent, E. P., *J. Chromatog.* **16**, 89–98 (1964).
29. Rutenberg, M. A., and Craig, L. C., *Biochemistry* **4**, 11–18 (1965).
30. Carnegie, P. R., *Nature* **206**, 1128–1130 (1965).
31. Porath, J., *Biochim. Biophys. Acta* **39**, 193–207 (1960).
32. Axen, R., Porath, J., and Ernback, S., *Nature* **214**, 1302–1304 (1967).
33. Daisley, K. W., *Nature* **191**, 868–869 (1961).
34. Saunders, D., and Pecsok, R. L., *Anal. Chem.* **40**, 44–49 (1968).
35. Ringertz, N. R., and Reichard, P., *Acta Chem. Scand.* **14**, 303–311 (1960).
36. Gelotte, D. B., and Emneus, I. A., *Chem.-Ingr.-Tech.* **38**, 445–451 (1966).
37. Crestfield, A. M., Moore, S., and Stein, W. H., *J. Biol. Chem.* **238**, 622–627 (1963).
38. Killander, J., Ponten, J., and Roden, L., *Nature* **192**, 182–183 (1961).
39. Greenwood, F. C., Hunter, W. M., and Glover, J. S., *Biochem. J.* **89**, 114–123 (1963).
40. Acred, P., Brown, D. M., Hardy, T. L., and Mansford, K. R. L., *Nature* **199**, 758–759 (1963).
41. Hummel, J. P., and Dreyer, W. J., *Biochim. Biophys. Acta* **63**, 532–534 (1962).
42. Herries, D. G., Bishop, W., and Richards, F. M., *J. Phys. Chem.* **68**, 1842–1852 (1964).
43. Roberts, G. P., *J. Chromatog.* **22**, 90–94 (1966).
44. Morris, C. J. O. R., *J. Chromatog.* **16**, 167–175 (1964).
45. Johansson, B. G., and Rymo, L., *Acta Chem. Scand.* **18**, 217–223 (1964).
46. Spahr, P. F., *J. Biol. Chem.* **239**, 3716–3726 (1964).
47. Tenenhouse, A., Arnaud, C., and Rasmussen, H., *Proc. Natl. Acad. Sci. U.S.* **53**, 818–822 (1965).
48. Potts, J. T., Jr., Reisfeld, R. A., Hirsch, P. F., and Munson, P. L., *Am. J. Med.* **43**, 662–667 (1967).
49. Putter, I., Kaczka, E. A., Harman, R. E., Rickes, E. L., Kempf, A. J., Chaiet, L., Rothrock, J. W., Wase, A. W., and Wolf, F. J., *J. Am. Chem. Soc.* **89**, 5301 (1967).
50. Rasmussen, H., and Craig, L. C., *Biochim. Biophys. Acta* **56**, 332–338 (1962).
51. Crestfield, A. M., Stein, W. H., and Moore, S., *J. Biol. Chem.* **238**, 618–621 (1963).
52. Heitz, W., and Coupek, J., *J. Chrom.* **36**, 290–301 (1968).

V *Ion Exchange*

I. GENERAL DISCUSSION

A. Introduction

1. Types of Exchange Reactions

Ion exchange reactions are reactions of a polyelectrolyte with any compound containing ionized groups which result in a change in the composition of the polyelectrolyte. Although these reactions may take place in solution and result in compounds or complexes of modified physical properties, this discussion will be concerned primarily with the use of insoluble polyelectrolytes (ion exchange substances) in separation processes. Ordinarily such processes include the steps of removing the desired compound from solution by contacting with an ion exchange substance and subsequent recovery from the ion exchange substance. The separation may be either group or fractionation type. In some cases ion exchange reactions can be useful without removing the compound from solution. Thus an ion exchange reaction for changing the salt form of the compound, such as conversion from chloride to sulfate or sodium to potassium, for the preparation of free base or free acid forms, or for removing excess salt may be the most practical method of accomplishing the desired step. In these cases the desired substance may or may not be removed from the process stream by the ion exchange material.

2. Ion Exchange Substances

Ion exchange materials useful in separation processes contain fixed charges in an insoluble solid matrix. These charges are neutralized by a mobile counterion which can exchange with other ions of like charge.

The fixed charges may be anionic or cationic and include groups with vastly different dissociation constants. The fixed charges are attached to a cross-linked polymeric backbone. With the exception of a few specialized inorganic compounds, such as zeolite, the polymeric backbone is an organic polymer. Of the many possible backbone polymers, the most commonly used are cross-linked polystyrene, cellulose, polyacrylates, and polyamines. A listing of commercially available types is contained in Appendix II.

B. Properties of Ion Exchange Resins

1. Physical Properties

Ion exchange resins are completely water-insoluble materials. Most modern synthetic resins are prepared in amorphous bead form. Some exchange materials, such as those based on cellulose, are fibrous and others are crystalline. A large range of particle sizes is usually available.

Usually the materials are physically stable and resist mechanical breakage. Some resins may shatter on drying or treatment with solvent. Some types are relatively hard and may be packed in deep beds without deformation; others, notably those used for large molecules, may be rather soft and easily deformed by pressure.

Volume changes caused by change of solvent or salt form may be slight for some resins but as much as twofold for others.

For any particular problem, optimization of the resin employed is influenced by both variations in the physical properties and the chemical properties discussed below.

2. Chemical Properties

a. Osmotic Forces Due to Charge Group Density. Consider a dry ion exchange particle containing fixed anionic sites associated with sodium ions. If such a particle is placed in pure water, water enters the particle and hydrates the charge sites and polymer backbone. Since the resin contains electrolyte internally an osmotic force tending to expand the resin particle is produced. If the polymer backbone is not completely rigid, the particle expands until the configuration is such that the osmotic force tending to further expand the particle is counterbalanced by the bonds forming the polymer backbone. Generally, the amount of water imbibed by the particle is related to the number of fixed charges per unit volume and the number of bonds per unit volume which form the three-dimensional network. Thus many loosely cross-linked resins expand more and have a greater distance between polymer chains and a greater volume per unit charge than tightly cross-linked resins. Some resins with rigid structures which do not expand appreciably on hydration have been recently introduced. In these resins also the internal osmotic pressure due to fixed charges is balanced by the restraining forces of the polymer network. For any *hydrated* resin an equilibrium is reached which can be defined in terms of water content per unit charge. As this value decreases the corresponding density of polymer and mobile ions within the particle increase. Highly cross-linked polymers with a low water content per unit charge and small average pore diameter do not undergo exchange reactions with large organic molecules readily. It has also been observed that large molecules permeating the polymer matrix are difficult to remove.

It is evident that the charge sites within the resin are inhomogeneous. Even if the charged groups are chemically identical and spaced evenly along a polymer backbone, e.g.,

$$-\underset{\substack{|\\X}}{CH}-CH_2-\underset{\substack{|\\X}}{CH}-CH_2-\underset{\substack{|\\X}}{CH}-CH_2-$$

the very fact that the polymer chains are linked in a three-dimensional network by some cross-link leads to differences in charge separation. Additional inhomogeneity is introduced by the fact that cross links are not uniformly spaced and the possibility that charged groups may not be chemically identical.

Sulfonated polystyrene resins have been studied extensively with regard to water content, degree of sulfonation, and cross-linking. If 2% divinylbenzene is incorporated, the average space between cross-links would be about 25 styrene units. Sulfonation of every available styrene unit (exclusive of cross-links) would produce a product with an effective equivalent weight of about 187. If 10% of divinylbenzene is added, the calculated equivalent weight

TABLE V.1

CAPACITY (%) OF POLYSTYRENE SULFONIC ACID RESIN FOR ORGANIC IONS

% Divinylbenzene	Organic compound[a]			
	I	II	III	IV
1	—	—	53	54
2	100	74	51	55
4	—	—	50	52
5	90	48	—	—
8	—	—	41	45
10	69	10	—	—
15	63	neg.	—	—
16	—	—	18	22
IR–120	—	—	40	46
IR–200	—	—	30	34
Amberlyst–15	—	—	32	34

[a] I = tetramethylammonium, eq. wt. 74 (*1*); II = cetyltrimethylammonium, eq. wt. 284; III = quinine, eq. wt. 324 (*2*); IV = cinchonine, eq. wt. 294.

would be 197. Actual values are about 188 and 194, respectively. The latter indicates some sulfonation of the cross-linking groups.

For the same resins, the water content of the polymer is about 80% and 52%, respectively. From this it can be calculated that the polymers contain 750 and 210 g of water per charge equivalent.

It is instructive to consider the situation that would exist within the resin particle if all of the exchange sites were occupied by a relatively large monovalent organic ion, such as cetyltrimethylammonium (eq. wt. 284). If the same amount of water were retained per equivalent (which obviously is not possible) the resulting polymers would contain about 380 mg and 1.35 g of organic material per gram of water. Since the internal space is limited by the restraining polymer network, the water content would necessarily decrease as the space is taken up by the large molecule. Osmotic forces tend to prevent water exodus and since this is greater with increased density of charge groups, it is apparent that the degree of exchange would be limited even if the pore size of the resin permitted free exchange.

Some data showing capacity for large organic ions are tabulated in Table V.1.

B. Exchange Equilibrium Constant. The exchange reaction is illustrated by the following general equation:

$$R^- \ldots Na^+ + M^+ \ldots X^- \rightleftharpoons R^- \ldots M^+ + Na^+ \ldots X^-$$

in which R^- represents the polymer containing anionic groups, and M^+ represents the counterion present in the solution phase. Assuming that the "activity" of X^- is the same regardless of the ratio M^+/Na^+ in the solution phase, the equilibrium at a certain M^+/Na^+ ratio can be formulated in terms of the mass action law as follows:

$$K_s = \frac{[R^- \ldots M^+](Na^+)}{[R^- \ldots Na^+](M^+)} \qquad (V.1)$$

This simple expression is useful in an empirical evaluation of a potential ion exchange process. However, correction to an expression satisfying thermodynamic requirements is complex and has occupied the attention of ion exchange chemists for the last 15 years.

Rather than dwell on the thermodynamic problems encountered, the purpose of this discussion will be to examine the interplay of the factors involved in the use of ion exchange reactions as a separation method. It is apparent that at some ratio of Na^+/M^+ in solution and at some total amount of M^+ relative to Na^+, a certain ratio of $R^- \ldots M^+/R^- \ldots Na^+$ would exist and that the ratio is probably different than 1. Therefore, the composition of the resin phase indicates "selectivity" for either Na^+ or M^+.

This selectivity represents the combined factors of osmotic pressure, polymer–substrate interaction, dissociation constants of the charge groups, and degree of exchange. For these reasons K, as formulated above, is not constant as the resin composition changes.

The selectivity coefficients Na^+/H^+ and Na^+/Li^+ have been rigidly calculated (*3*) using partial molal volumes and activity coefficients derived from nearly completely expanded polymer. Although these calculations are remarkable close to the observed values, the data required for the calculation are so extensive that the procedure is of little practical value in most cases.

Since the resin phase has a fixed number of charged groups, or exchange capacity, this leads to selectivity or preference for ions of higher valence at low solution concentration. Expressing the composition of the resin and solution phases as fractions of the total composition, i.e.,

$$F_r^{Na^+} = \frac{[R^- \ldots Na^+]}{C_r} \quad \text{and} \quad F^{Na^+} = \frac{(Na^+)}{C}$$

in which C_r is the total resin capacity and C the total solution concentration, the following relation is obtained:

$$K_s = \frac{F(1 - F_r)}{F_r(1 - F)} \tag{V.2}$$

If, however, a valence difference exists between the exchanging ions, and assuming again that the resin phase is homogeneous, K_s is formulated as follows:

$$K_{M_2/M_1} = \frac{[R \ldots M_2]^x(M_1)^y}{[R \ldots M_1]^y(M_2)^x} \tag{V.3}$$

in which x is the valence of M_1 and y is the valence of M_2 in the exchange reaction.

The fractional expression can then be written as

$$K_{M_2/M_1} = \left(\frac{C}{C_r}\right)^{y-x}\left(\frac{F}{F_r}\right)^y\left(\frac{1 - F_r}{1 - F}\right)^x \tag{V.4}$$

Since C_r is the concentration in the resin phase and is constant except for a small difference due to swelling or contracting, the selectivity is exponential with the difference in valence of the two compounds at constant solution concentration.

The ion exchange properties of multivalent basic aminoglycoside antibiotics reflect this general property. Early studies with streptomycin (see Chapter III, Section V) indicated ready removal by cation ion exchange resins of the antibiotic from dilute process streams, such as fermentation broth, in the presence of much greater concentration of inorganic ions.

Attempts to recover the antibiotic by elution from the resin with salt solutions gave uniformly poor results, in that very large amounts of salt were required. Since the antibiotic contains strongly basic guanido groups and is unstable at high pH it could not be recovered by elution at high pH. It was not until a study of carboxylic ion exchange resins led to the development of 1RC–50 that a satisfactory ion exchange recovery procedure for streptomycin became available.

c. DONNAN MEMBRANE POTENTIAL. An ion exchange particle can be likened to a dialysis cell containing a membrane permeable to the exchanging ions and their mobile co-ions but not to a polyelectrolyte. Such a configuration can be depicted as

$$M^+, Na^+, X^-/\text{membrane}/M^+, Na^+, X^-, R^-$$

in which R^- represents the polyelectrolyte.

At equilibrium the chemical potential of the ions M^+ and Na^+ is equal on both sides of the membrane and the requirement for electroneutrality causes X^- to be excluded from the polyelectrolyte solution. This effect is known as the Donnan membrane potential.

For example, if a polyelectrolyte solution containing x equivalents of Na^+ per liter is placed on one side of a membrane and a solution of NaCl containing y equivalents per liter is placed on the other side and $x > y$, Na^+ would have a diffusion tendency from the polyelectrolyte to the NaCl solution and Cl^- would have a diffusion tendency into the polyelectrolyte solution. Both diffusion tendencies cause an electrical imbalance. That is, the NaCl solution would become positively charged due to loss of negative ions and gain of positive ions whereas the polyelectrolyte solution would become negatively charged due to loss of positive ions and gain of negative ions. The resulting electrical force creates a potential which reverses the diffusion tendency. At equilibrium the Donnan membrane potential balances the diffusion tendency. The greater the difference between x and y the greater the membrane potential. Hence, the exclusion of the co-ion from the resin phase increases with increasing density of charged groups in the resin.

D. DISSOCIATION CONSTANT AND HYDROLYSIS. Water surrounding the ion pair within the resin particle can hydrolyze the complex. This is illustrated as follows for a resin with fixed anionic charges (cation exchange resin):

$$(RM) \rightleftharpoons R^-M^+ + HOH \rightleftharpoons R^-H^+ + M^+OH^-$$

$$K_{d,r} = \frac{[R^- M^+]}{(RM)}$$

which in turn leads to the establishment of the following equilibria

$$R^-H^+ \rightleftharpoons (RH)$$

$$K_{d,r} = \frac{[R^-][H^+]}{(RH)}$$

$$M^+ OH^- \rightleftharpoons (MOH)$$

$$K_{BOH} = \frac{[M^+][OH^-]}{(MOH)} = \frac{[M^+]\,10^{-14}}{(MOH)[H^+]}$$

In the above expressions the quantities in parentheses ( ) are used to represent nonionized forms, as a method of indicating qualitatively that such forms may exist. Thus the stability of the resin complex is influenced by the dissociation constants of the fixed charge and the mobile charge. Since the charge per unit volume inside the resin particle is high, an appreciable percentage of the resin functional groups may be in the nonionized state. This is especially true if the functional group is a weak acid or base. Similarly the dissociation constant of the counterion and its potential co-ion in the solution phase influence the stability of the complex. For cation exchange the percentage of dissociation of the resin functional group is increased by increasing pH whereas the percentage of dissociation of the counterion is decreased by increasing pH, and it clearly may be impossible to achieve a stable complex at any pH if both $K_{d,r}$ and K_{BOH} are small. This would be the case with an ion exchange resin with weak acid functions and exchange reactions with weak bases. The reverse applies to anion exchange.

The actual measurement of the effective dissociation constant of either the functional group or the counterion within the resin is impossible. However, the extremely small leakage of ions from strongly ionized exchange resins into dilute solution has sometimes been interpreted as an indication of a small K_{rm}.

In addition, because of the factors contributing to inhomogeneity mentioned above, the dissociation constant is usually not equal for all exchange groups within the resin. Anion exchange resins are frequently less homogeneous than cation exchange resins and an additional factor, nonidentity of the ionizing groups, is frequently present. By titration in the presence of electrolyte the approximate pK values of a number of functional groups have been determined. These are summarized in Table V.2 and illustrated in Figs. V.1 and V.2.

A detailed discussion on the pK of functional groups in the resin matrix and methods of measurement is given by Helfferich (*4*).

E. POLYMER–SUBSTRATE INTERACTION. In some cases the organic polymers used in ion exchange resins appear to exhibit solvent properties. Thus, anion

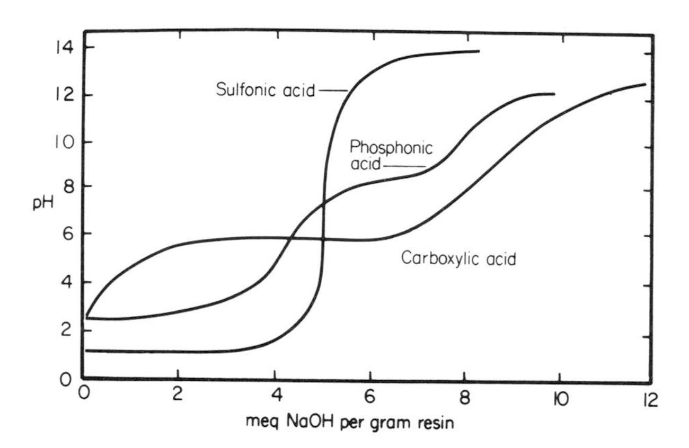

FIG. V.1. Titration curves of typical cation exchange resins.

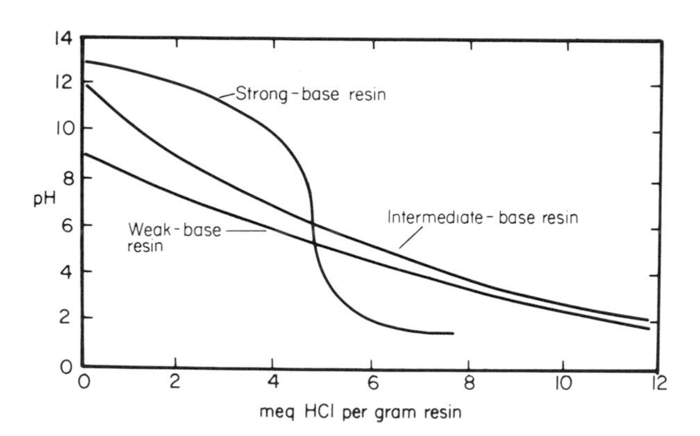

FIG. V.2. Titration curves of typical anion exchange resins.

TABLE V.2

APPARENT pK VALUES FOR RESIN FUNCTIONAL GROUPS

Group	pK_a
$-SO_3H$	<1
$-PO_3H_2$	pK_1 2–3, pK_2 7–8
$-COOH$	4–6
$-N(CH_3)_3{}^+$	>13
$-N(CH_3)_2CH_2CH_2OH^+$	>13
$-NR_2$	7–9
–N– R	7–9

exchange resins may remove more than 1 equivalent of acetic acid from aqueous solution. Apparently neutral compounds (e.g., vitamin B_{12}) are adsorbed by carboxylic acid resins in the acidic form but not in the salt forms (*5*). The separation of polar basic antibiotics on anion exchange resins in the free (hydroxyl) form appears to involve forces similar to solvent extraction (*6*). Organic acids are retained by sulfonic acid resins and, on elution with water, appear in the eluate in order of decreasing polarity as though extracted by the resin phase (*7*). Some large nonpolar organic acids are readily adsorbed by anion resins and cannot be removed by aqueous eluents. These materials may be removed in the presence of solvents. Compounds which are recovered with difficulty, if at all, from polystyrene-based resins can sometimes be processed with resins based on hydrophilic polymeric backbone, such as cellulose or Sephadex.

The formation of hydrophobic bonds has been postulated to account for these observations as well as the very high selectivity coefficients observed with many large organic compounds. The electrostatic attraction due to the ionized groups may bring the organic compound into close contact with the polymer backbone. The elimination of water between the polymer and adsorbed organic compound would then produce a complex having little tendency to dissociate, since this would involve redissolving the adsorbed substance in the water phase.

It should be stressed that the formation both of an ionic and a hydrophobic bond between the adsorbed substance and the polymer is always less reversible than ion exchange alone. In the latter case, the mobile ion exerts little effect on the molecular arrangement of the polymer backbone and can retain a high degree of mobility within the matrix. As a result, the distribution of such an ion is probably almost uniform throughout the ion exchange matrix at any degree of saturation.

The tendency of organic compounds to form "immobile" complexes results in turn in preferential loading near the surface of the resin particle. In addition, site "saturation" of the potential hydrophobic bonding sites may occur prior to saturation of exchange sites.

The drastically reduced mobility of adsorbed molecules and the requirement for diffusion of unbonded molecules past already adsorbed molecules results in the necessity for slow flow rates in the treatment of organic molecules. Although chromatography of organic compounds on ion exchange resins is highly successful in many cases, rapidly changing distribution coefficients as the loading is increased necessitate the use of relatively low loading. Usually less than 5% of the total available ion exchange sites are occupied by substrate. Many successful procedures have been developed using a gradient elution method to compensate for the changing distribution coefficient.

If the organic compound forms hydrophobic bonds with the polymer backbone which are not readily solvated by water or the solvent permeating the resin, it is apparent that "detachment" of the ionic complex by high concentrations of competing ions or by changing the pH will not be efficient. Under this circumstance elution will be favored by using solvents which can "solvate" either the polymer backbone or the adsorbed substance.

For many compounds aqueous solutions of lower alcohols have been found useful in the elution step (*8*, *9*). Since some salt is usually also necessary, the amount of solvent required for maximum efficiency depends upon the individual compound. In general, higher solvent concentrations (which may not permit simultaneous use of high concentrations of salts) are required as the polarity of the adsorbed compound decreases. Combinations of methanol–water with salts such as ammonium chloride, potassium acetate, or iodide are frequently useful.

F. INFLUENCE OF CO-ION ON THE EXCHANGE PROCESS. The formation of the resin complex is also influenced by the co-ion in solution even though the ion does not take place in the exchange reaction. At equilibrium the "activity" of a substance in the solvent in the resin phase equals that in the solution phase, providing an exchangeable counterion is present. Thus, if γ_r is activity in the resin phase and γ_s activity in the exterior solution phase, then $\gamma_r = \gamma_s$,

$$\gamma_s = C_s f_s$$

and

$$\gamma_r = C_r f_r$$

in which f is the fraction of the total contributing to activity. For many organic compounds, rigid mathematical treatment of f for either phase has not been carried out as yet.* In the solution phase, the dissociation constant from the mass action expression,

$$K_a = \frac{(H^+)(A^-)}{HA} \quad \text{or} \quad K_b = \frac{(B^+)(OH^-)}{BOH}$$

reflects the effect of pH on the substrate.

For nonpolar organic compounds, association into aggregates, perhaps

* This simple expression does not provide any insight into the factors which affect f_r. These are multiple in nature and must include pressure effects on the gel network, ion pair formation, and the effect of competing ions. Considering an ion exchange reaction of the type, $R^-A^+ + B^+ \rightleftharpoons R^-B^+ + A^+$ under the circumstances in which f_s for A^+ and B^+ are identical, then selectivity becomes merely the ratio $f_{r,B}/f_{r,A}$. Hence an observed selectivity ratio different than unity is only an indication that for the system employed $f_{r,A} \neq f_{r,B}$. If $f_{r,B}$ is lower than $f_{r,A}$, $(\gamma_r)_B$ and $(\gamma_s)_B$ are also low and the resin would "prefer" B.

even micelles, may depress solution activity. This reaction may be influenced by the co-ion present.

Since both counterions are influenced by the nature and concentration of the co-ion, the contributions of these terms vary as the solution composition changes. This may result from the ion exchange process itself. An illustration of this effect would be the exchange of calcium ascorbate with resin citrate. Although calcium does not take part in the exchange reaction, sequestering of citrate would provide additional driving force for the exchange by influencing f_s. It is obvious that if a mixture of cations including calcium were present, some of this effect would be obtained, albeit diminished.

For many ion exchange processes complete analysis of the system is not possible. For applications in fractionation systems the dissociation properties of the resin and substrate are partially controlled by using sufficient buffer to "swamp" effects due to minor concentrations of different counterions and to prevent much pH drift. To aid in maintaining nearly constant ionic conditions the resin should be thoroughly conditioned with the buffer to be used. Complex buffers consisting of mixtures of salts are undesirable and should be avoided if possible.

In group separation procedures, the single most important variable is usually the dissociation constant, since this allows rough fractionation as mentioned above. However, yield limitations and overlap of group types may occur due to the influence of many factors on the overall f of either the solution or resin. Thus, as the concentration of the substrate diminishes the overall f_s would increase, resulting in increased driving force into the resin phase. Although this effect is of great value in the removal, by ion exchange, of substances present in very low concentration, it may prevent suitable recovery in the elution stage.

G. Resin Capacity and Overall Distribution Coefficient. The overall distribution coefficient, G, is usually expressed as a function of concentration in the resin and solution phases, i.e.,

$$G = C_s/C_r$$

The relationship of C_r and C_s is described by K_s, as previously mentioned, and is influenced by other ions in solution. Since C_r is limited by the number of active sites in the resin (capacity) or by the total available space for large organic molecules, G is nonlinear with concentration and approaches the saturation point asymptotically.

In chromatographic separations in which it is desirable to maintain G constant or nearly so, only small resin loadings are employed.

If the resin is used in columns for group separation, high loading is usually desirable. A plot of concentration of the material in the spent solution

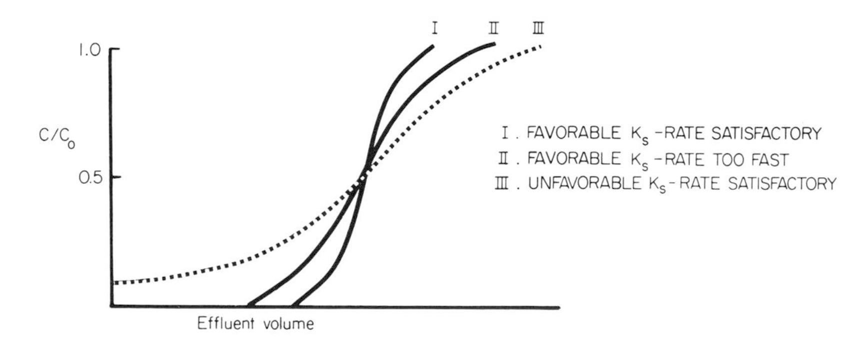

FIG. V.3. Typical breakthrough curves.

against time at constant flow rate gives an S-shaped curve. The position and shape of the curve are influenced by the fundamental selectivity coefficients and flow rate. Typical breakthrough curves as illustrated in Fig. V.3.

The curves depicted in Fig. V.3 are somewhat idealized but define the conditions for good performance. Three general curves are depicted. In the first, good K_s, a relatively sharp front is produced and there is not much increase in column loading once breakthrough has started. This represents the ideal situation. If the K_s is lower a flat curve may be obtained and it is difficult to obtain high yields because a large percentage of column capacity can only be obtained with substantial losses in the spent stream.

Adsorption kinetics also influence the shape of the breakthrough curve. As the flow rate is increased the curve becomes shallower even if K_s is satisfactory.

In summary, the distribution coefficient, G, is usually not constant and is influenced by the following variables:

(a) The concentration of competing ion or ions, i.e., selectivity coefficient
(b) The degree of dissociation of the resin
(c) The degree of dissociation of the substrate
(d) The degree of saturation of the resin with the substrate
(e) The formation of hydrophobic bonds between polymer backbone and organic substrate
(f) Swelling or contraction of resin phase

C. Resin Types

A variety of "synthetic" ion exchange resins is available. Although inorganic compounds, in particular certain zeolites and zirconium-based ion exchangers, are known, these have found less use than substances based on organic polymers. The chemical activity of an ion exchange resin can be varied by changing

(1) The type of polymer used

(2) The density of polymer, i.e., the degree of cross-linking between polymer chains

(3) The type of functional group, i.e., strong acid, strong base, weak acid, weak base

(4) The density of ionizable groups within the resin matrix

(5) Particle size; fine mesh resins have a higher ratio of surface to internal groups

(6) Inhomogeneity of the resin due to nonuniform distribution of functional groups or the presence of more than one type of functionality

Although several hundred different types of ion exchange resins have been made available by different manufacturers, these are based mostly on polystyrene, polyacrylic acid, polycarbohydrate, or polyamine structures. A general review of these types follows and a list of various types currently available is contained in Appendix II.

1. Strong Acid Resins

Polystyrene-based resins are most commonly utilized for large-scale applications, especially water treatment. Usually the polymerization is carried out prior to activation, usually by sulfonation. The resulting polymer contains almost one sulfonic acid group for each benzene nucleus. Coal and phenol–formaldehyde polymers can be sulfonated to yield strong acid resins. Cellulose or dextran polymers can be treated to give a low yield of sulfate ester suitable for some types of ion exchange. Polymers of vinylsulfonic acid have been prepared but are not commercially available. Styrene-based resins

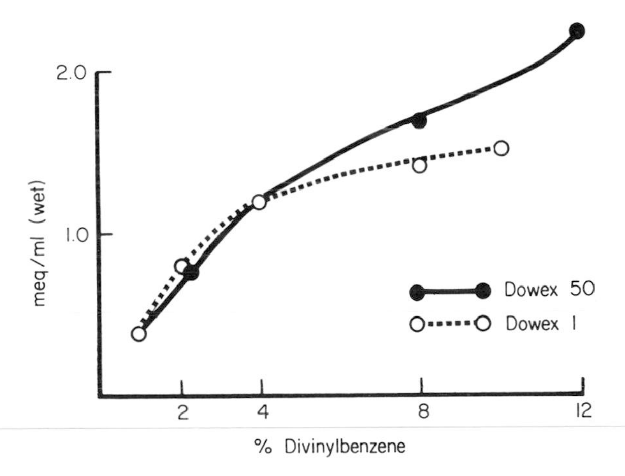

Fig. V.4. Effect of cross-linking agent on volume capacity of ion exchange resins.

carrying a large alkyl group as well as a sulfonic group have been prepared (*10*) and are useful in nonaqueous media.

For polystyrene-based resins the degree of cross-linking is controlled during polymerization by the fraction of cross-linking agent, divinylbenzene, present in the mixture. After sulfonation under equivalent conditions and conversion to the sodium form, the water content increases with decreasing cross-linking. The number of exchange groups per unit volume of water-swollen resin is indicated in Fig. V.4 for both sulfonated polystyrene (Dowex 50) and a quaternized anion exchange resin (Dowex 1).

2. *Weak Acid Resins*

Weak acid resins contain carboxyl functions which may be attached to a cross-linked hydrocarbon chain (acrylate or methacrylate).

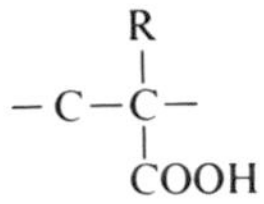

$$-C-\underset{\displaystyle COOH}{\overset{\displaystyle R}{\underset{|}{\overset{|}{C}}}}-$$

Treatment of carbohydrate (cellulose or dextran) under alkaline conditions with chloroacetic acid yields a product containing a carboxymethyl group $-CH_2COOH$. Using cellulose the resultant product, carboxymethyl cellulose, has found extensive use in the purification of large molecules, especially proteins. The degree of substitution can be varied somewhat. Controlled oxidation of cellulose yields oxycellulose which contains carboxyl groups attached to the cellulose chain. Variations in the pK of the carboxyl function (as much as 1 pH unit) which are obtained with different polymer types may be of critical significance for some fractionation processes. Polymers in which the –COOH is attached directly to an aromatic nucleus or containing phenoxyacetic acid moieties have also been prepared.

3. *Intermediate Strength Acid Resins*

Although of limited utility phosphate and phosphonic acid resins are available. The exchange groups may be attached to carbohydrate chains or to aromatic polymers.

4. *Strong Base Resins*

One of the most useful strong base resins is obtained from polystyrene by chloromethylation and subsequent amination.

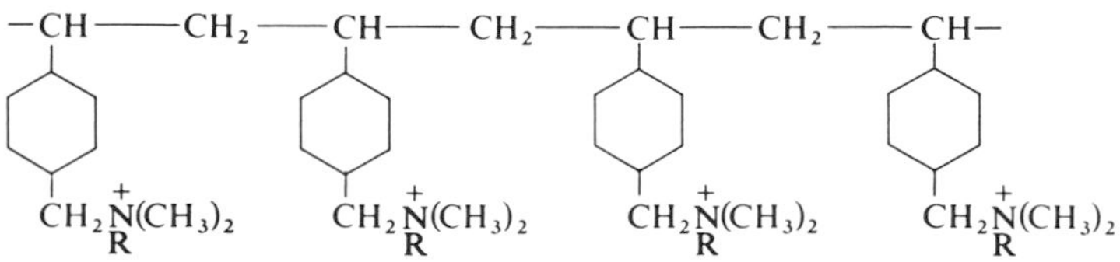

$$-CH-CH_2-CH-CH_2-CH-CH_2-CH-$$

(each CH bears a benzene ring carrying $CH_2\overset{+}{\underset{R}{N}}(CH_3)_2$)

Type I ($R = CH_3$) is obtained using trimethylamine as the quaternizing agent. Type II resin ($R = CH_2CH_2OH$) is obtained by using dimethylethanolamine. The corresponding quaternary resin has slightly weaker base strength than Type I. Since the $-CH_2Cl$ grouping is capable of reacting with a neighboring aromatic nucleus, secondary cross-linking and resultant loss of capacity may occur during the resin preparation.

Many other quaternizing agents are possible and may be useful for special purposes. Other polymer types have found occasional usage. Among these are polymers derived from vinylpyridine, vinylimidazole, and polyethyleneimine.

5. *Weak Base Resins*

By using a primary or secondary amine as the aminating agent, anion exchange resins having primarily only secondary or tertiary amine functions may be obtained. In this case also the presence of the reactive $-CH_2Cl$ grouping leads to secondary reactions which include quaternization. Weak base resins are also obtained by esterification of polycarboxyl-containing polymers with amines such as ethanolamine or substituted derivatives. The reaction of β-chloroethyldiethylamine with alkaline cellulose results in the formation of polymer containing $-OCH_2CH_2N(C_2H_5)_2$ groups. The resulting product, DEAE cellulose, has found extensive use in the purification of proteins and nucleotides. The group may be added to dextran polymers.

6. *Intermediate Base Strength Resins*

Polymers obtained by treating guanidine with formaldehyde are stronger bases than tertiary amines but are of weaker base strength than quaternary ammonium resins. Guanido groups can be introduced into other polymers. Resins containing base groups of varying base strength are obtained from cellulose by treatment with the corresponding activated amine.

D. Resin Use

1. *Resin Cycle*

In most applications ion exchange resins are regenerated and reused. Applications based on irreversible processes are relatively rare. Thus the steps in an ion exchange process consist of

(1) *Resin conditioning,* i.e., conversion to the counterion state, hydrogen, sodium, etc., to be used in the treatment step

(2) *Resin loading,* treatment with the material being adsorbed or chromatographed

(3) *Resin elution,* elution of desired material either chromatographically or as a mass displacement

In some cases the eluted resin may be sufficiently conditioned for reuse without a separate conditioning step. In other cases the conditioning treatment may be extensive. In column operations the flow rate may be limited by either pressure drop in the equipment or diffusion of the material being processed. For the concentration of large organic ions which are present in the treated stream in low concentration a complete resin cycle may require as long as 14 days.

2. *Ion Exchange Reactions*

For resin loading or unloading (elution) two general types of reaction may be used:

(1) *Exchange*

$$RX + AB \rightleftharpoons RA + BX$$

X is a salt form of the resin.

(2) *Neutralization*

$$RH(OH) + AOH(H) \rightleftharpoons RA + H_2O$$

For loading the reaction is carried out in the forward direction; elution is the reverse. Two additional reactions are of importance in the elution step:

(3) *Repression of resin ionization*

$$RA + H^+(OH^-) \rightarrow RH(OH) + A^+(A^-)$$

(4) *Repression of substrate ionization*

$$RA + H^+ \rightarrow R^-H^+ + HA$$
$$RB + OH^- \rightarrow R^+OH^- + BOH$$

In type (3), strongly ionized substances are removed from weakly ionized resins by depressing the ionization of the resin. The procedure is used for the recovery of streptomycin, which contains strongly ionized guanido groups, from carboxyl-containing resins by treatment with acid.

The procedure illustrated by reaction type (4) is of great value for the elution of weakly ionized substances from strongly ionized resins. Thus, amines can be eluted from sulfonic acid resins by treatment with bases or carboxylic acids can be eluted from strongly basic anion exchange resins by treatment with acid.

E. Group Separations

In considering an ion exchange step for an organic compound the most important property is the p*K* or the influence of pH on ionization potential of the compound. Of nearly equal importance is the size or shape of the molecule. Finally, solubility in aqueous media may significantly affect the adsorption coefficient. Stability, as affected by pH, is also very important.

Ion exchange processes are admirably suited for group separation since the resin process may be manipulated to group compounds based on their pK values. Since the latter can be obtained from either paper chromatography or electrophoresis data (Chapter II) conditions for a particular compound can be specified. The Henderson-Hasselbach relation, $\mathrm{pH} - \mathrm{p}K = \log$ (ion/non-ionized), is useful in determining the pH to be used for adsorption and elution steps. In the absence of actual data the equation can be useful in predicting the properties of the ion exchange resins used.

A general procedure for the separation of four main groups, strong acids or bases and weak acids or bases is as follows:

1. Separation of Strong Bases

A. NO STABILITY LIMITATIONS. Selectivity and removal of weaker bases can be obtained by carrying out the adsorption at high pH (either strong or weak acid resin can be used) or by carrying out the adsorption at a low pH and washing the resin adsorbate with high pH buffers. Weak bases can be preferentially removed from the resin. Either procedure yields a resin adsorbate containing only strong bases. These can then be removed from the resin by application of reaction (1) using an appropriate salt or by reaction (3) if a weak acid resin has been used.

B. UNSTABLE AT ALKALINE pH; STABLE AT NEUTRAL AND ACIDIC pH. Under these conditions exclusion of bases of intermediate pK may be impossible. The adsorption would be carried out using a weak acid resin (carboxylic) at the highest pH compatible with the stability. The elution is by acidification with a strong acid.

C. UNSTABLE AT ACIDIC pH; STABLE AT NEUTRAL AND ALKALINE pH. Adsorption conditions would be the same as (A) with no limitations on resin type. Elution is carried out with salt using a cation preferred by the resin.

D. STABLE ONLY AT NEUTRAL pH. Similar to (B) except elution with salt must be used.

E. STABLE ONLY AT ACIDIC pH. Separation from weak bases is not possible. Adsorption on strong acid resin followed by salt elution yields a mixture of all types of bases.

F. STABLE ONLY AT ALKALINE pH. Adsorb and elute with salt at high pH.

2. Separation of Strong Acids

A. NO STABILITY LIMITATION. Adsorb on either strong or weak base resin at low pH. Elute strong base resin with salt or weak base resin with alkali.

B. UNSTABLE AT ALKALINE pH; STABLE AT NEUTRAL AND ACIDIC pH. Carry out adsorption on strong base resin at the minimum pH compatible with stability. Elute with appropriate salt. If the pH range is satisfactory use a weak base resin and elute with alkali.

C. UNSTABLE AT ACIDIC pH; STABLE AT NEUTRAL AND ALKALINE pH. Adsorb at low pH. Elute with appropriate salt.

D. STABLE ONLY AT NEUTRAL pH. A separation based on pK is not possible but a mixture of acids can be obtained by adsorption on strong base resin and elution with salt.

E. STABLE ONLY AT ACIDIC pH. If there is no lower pH limit on stability the adsorption can be carried out at low pH on either strong or weak base resins. The elution is carried out with strong acid or salt.

F. STABLE ONLY AT ALKALINE pH. The separation based on pK is not possible. Acids can be obtained by adsorption on strong base resin followed by elution with salt. Intermediate strength resin might be eluted with alkali.

3. Separation of Weak Bases

A. NO STABILITY LIMITATION. Adsorb on strong acid resin; elute with weak base. Strong bases remain on resin.

B. UNSTABLE AT ALKALINE pH; STABLE AT NEUTRAL AND ACIDIC pH. Unless the compound is stable at about 2 pH units below pK_B, ion exchange separation is unlikely to succeed. If adsorption step at acidic pH can be carried out, the weak bases can be separated by elution with a weak base.

C. UNSTABLE AT ACIDIC pH; STABLE AT NEUTRAL AND ALKALINE pH. Separation from strong bases may not be possible. May be adsorbed at acidic pH and eluted with salt.

D. STABLE ONLY AT NEUTRAL pH. Separation from neutral compounds may not be possible. Strong bases are removed by adsorption on strong acid resin. Acidic compounds removed by adsorption on strong base resin.

E. STABLE ONLY AT ACIDIC pH. Separation from strong bases is not possible. Bases are separated by adsorption on strong acid resin and elution with salt.

F. STABLE ONLY AT ALKALINE pH. Acids and strong bases removed by adsorption with appropriate resins. Neutral compounds and weak bases remain.

4. *Separation of Weak Acids*

A. NO STABILITY LIMITATION. Adsorb on strong base resin at alkaline pH; elute with weak acid. Strong acids remain on the resin.

B. UNSTABLE AT ALKALINE pH; STABLE AT NEUTRAL AND ACIDIC pH. Separation from strong acids may not be possible. Adsorb acids at alkaline pH and elute with salt.

C. UNSTABLE AT ACIDIC pH; STABLE AT NEUTRAL AND ALKALINE pH. Separation from neutral compounds may not be possible. Strong acids and all bases are removed by appropriate ion exchange treatment.

D. STABLE ONLY AT NEUTRAL pH. Strong acids and bases removed by appropriate ion exchange treatment. Weak bases and neutral compounds may not be separated.

E. STABLE ONLY AT ACIDIC pH. Weak acids and neutral compounds remain after ion exchange removal of strong acids and all bases.

F. STABLE ONLY AT ALKALINE pH. Separation from strong acids may not be possible. Acids are removed by strong base resin and eluted with salt.

The above brief outline may be conditioned by factors other than ionization which influence the selectivity coefficient and differences in behavior of seemingly similar resins may be observed.

II. ION EXCHANGE APPLICATIONS

A complete bibliography of the applications of ion exchange would encompass practically all fields of modern separation chemistry, extending from separations of isotopes [e.g., $^{14}NH_4^+$ and $^{15}NH_4^+$] to the preparation of pure t-RNA (m. wt. 25,000). Separations may be based on ion exchange reactions or on adsorption or exclusion properties in which neutral compounds are separated.

The scale of operation varies from analytical separations with a few milliliters of resin (or even a single ion exchange bead) and microgram quantities of material to commercial production units containing several cubic meters of resin and producing tons of material.

The only common feature of the applications is that a two-phase system is used. The application method can be batchwise, semicontinuous batch–slurry, columnwise, or continuous column operation.

Although a major portion of the ion exchange substances manufactured is used in water conditioning, the following discussion will be limited to the use of ion exchange methods for the preparation of pure organic compounds, especially those materials produced from natural sources.

As in the other sections, two general types of separation, group and fractionation, will be considered. It is probably possible to use ion exchange substances to effect any separation so long as the materials being separated permeate the resin matrix (even this may not always be required). Since resins may be used in almost any solvent, partition effects other than ion exchange may be obtained. Common hexoses may be separated by this mechanism. The separation may be based on exclusion rather than adsorption, or on some property such as ligand formation, or partition between solvents of dissimilar composition within and without the polymer network.

The applications discussed below have been selected as typical of the general areas of application and are intended to provide initial guidance in the selection of experimental conditions and resins. The emphasis is based primarily on an experimental approach since no unifying theory of ion exchange behavior can be applied to the diverse types of resins and substrates and the complex mixtures encountered.

A. Separations Based on Ion Exchange Reactions

1. Group Separations

A. SALT REMOVAL—NEUTRAL COMPOUNDS. The removal of salt is frequently an important step in the preparation of pure compounds. Although gel filtration can be used in certain cases, as described in Chapter IV, deionization by ion exchange is more useful when large volumes of solution must be processed and is especially applicable to neutral compounds which are not retarded or adsorbed by the resin.

Elimination of salt requires both anion and cation exchange resins. For use the resins are converted to the free base and free acid form. The resins may be used in successive treatments in different columns or in mixed-bed columns. If the resins are used in sequence a resin with a large dissociation constant is required in the first cycle and all of the salt is converted to the corresponding hydroxide or acid by one of the following reactions:

$$R^-H^+ + Na^+Cl^- \rightarrow R^-Na^+ + H^+Cl^-$$

or

$$R^+OH^- + Na^+Cl^- \rightarrow R^+Cl^- + Na^+OH^-$$

In order for the exchange to function the ionization of the resin must not be appreciably depressed by the acid or alkali produced; therefore solutions with high salt concentrations may require more than one stage. In addition the compounds being treated must be stable at the pH produced since a finite time elapses before the treated solution is neutralized by the resin in the next stage. The ready hydrolysis of sucrose has always limited the use of ion exchange deionization in sugar refining although many attempts have been made to develop a satisfactory procedure.

Prevention of drastic pH changes and more complete deionization result if the two resin types are mixed in the same bed. Exhausted resin may be discarded (in small-scale applications) or regenerated after hydraulic separation in the tower. Operation of the latter procedure requires the use of resins with sufficiently different densities to allow separation when the resin bed is fluidized by high velocity upflow treatment.

For large-scale use the efficiency of utilization of regenerants and regeneration time may be important considerations. An early example of increased efficiency resulting from resin structure was the introduction of Type II quaternary exchange resins. These resins, which are based on a styrene polymer, are quaternized using dimethylethanolamine. Although there is little difference in total deionization capacity, in comparison with Type I resins, much greater efficiency of utilization of hydroxide is obtained with Type II.

Recently a resin with low affinity for HCO_3^- has been introduced, Rohm & Haas, Amberlite IRA68. With this resin a three-column system for desalting has been devised which uses only ammonia and sulfuric acid regenerants in high efficiency. The following reactions are used:

Col. 1

$$RNH^+HCO_3^- + Na^+Cl^- \rightarrow RNH^+Cl^- + Na^+HCO_3^-$$

Col. 2

$$RCOOH + Na^+HCO_3^- \rightarrow RCOO^-Na^+ + HCO_3^- + H^+$$

Col. 3

$$RNH^+ + HCO_3^- \rightarrow RNH^+HCO_3^-$$

Column 1 is regenerated with NH_3 and then used as Column 3 and Column 2 is regenerated with H_2SO_4.

B. Demineralization by Adsorption–Elution. If the product being sought adsorbs on ion exchange resin, complete ion removal as described above usually results in a yield loss. In this case demineralization may be possible by adsorption on the resin and elution with a volatile buffer. If the substance is stable under the adsorption–elution conditions this process may be quite advantageous if relatively small amounts of the desired compound are present compared to the amount of mineral to be removed. Some group separation may also be obtained in the same step. The reaction mixture produced by the acid hydrolysis of protein is such a mixture. The acid solution can be contacted with strong acid resin, which removes the amino acids allowing residual acid (or salt if the mixture has been partially neutralized) to pass into the effluent. The amino acids are recovered by contacting the resin with dilute ammonia. Evaporation of the ammonia eluate produces an essentially salt-free solution of amino acids. Good yields of acidic and neutral amino acids with the exception of the aromatic phenylalanine and tyrosine are obtained. The strongly basic arginine may be only partially recovered.

c. USE OF ION RETARDATION RESINS. Neutral compounds may be demineralized by the use of ion retardation resins. Ion retardation resins are specially formulated resins containing both acidic and basic charge groups in approximately equal concentration.

A commercially available type is prepared by adsorbing acrylic acid in a quaternary resin and then polymerizing the acrylic acid. Since the long-chain linear polymer is retained within the quaternary resin, such resins have been called "snake cage" resins (*11*).

These resins may be considered similar to a mixed bed except that the two types of exchanger are contained within the same polymer bead. In addition the reaction is reversed by merely washing with water.

By considering that the surface of the bead acts as a semipermeable membrane and remembering that both cationic and anionic charges are fixed on the inside of the resin, it can be seen that the Donnan potential is quite low since an ion migrating into the resin would liberate either a proton or a hydroxide ion from the fixed polyelectrolyte. The liberated ion can produce water by reaction with a neighboring group. This leaves a net charge inside the resin which attracts the opposite ion. Thus large amounts of electrolyte can permeate the resin particle without producing an electrical imbalance. At equilibrium the activity products on both sides of the membrane are equal and hence salt would move into a resin particle which was initially salt-free. If the salt solution is now replaced with water at pH 7, hydroxide ions and protons again redistribute and salt is ejected from the resin. Thus, if a charge containing both salt and a neutral compound is placed on a column, the neutral compound appears in the effluent prior to the salt.

The distribution term used for gel filtration can be applied to this system. Neutral compounds have a K_d of 0 and the K_d of the electrolyte varies with the resin type and electrolyte employed.

However, since the water content inside the resin is of less mechanistic importance it is more convenient to use a distribution term, K_{av}, to denote the distribution between the mobile phase and the stationary phase without regard to water content of the stationary phase

$$K_{av} = \frac{V_e - V_0}{V_s} \qquad \text{(V.5)}$$

where V_s is volume of stationary phase.

With 11A8 resin (Dow Chemical Company), the elution volume of sodium chloride is about 1.4 bed volume and K_{av} is then approximately 1.6. Since divalent ions are retained more strongly by the resin for the reasons outlined above, the observed K_{av} of $(NH_4)_2SO_4$ is about 9.0. Since ΔK_{av} is related to ΔK_d by a constant describing the solvent content of the stationary phase, expressions for resolution, charge volume, etc., for gel filtration (Chapter IV) may be used in ion retardation.

D. ION EXCLUSION. The tendency of ion exchange resins to exclude electrolytes based on Donnan potential has already been described. Since no such exclusion exists for neutral compounds, neutral and ionic materials may be separated since electrolytes appear in the effluent at an earlier volume than neutral compounds. This phenomenon has been designated ion exclusion by Wheaton and Bauman (*12*).

Neutral compounds may be retarded merely because of permeation of the gel phase (gel filtration) or because of backbone–substrate interaction. In either case the degree of separation is exactly analogous to that already calculated for gel-permeation chromatography if suitable K_d's are used. In fact the term K_d was introduced by Wheaton and Bauman.

Since the desalting is due to the ability of neutral compounds to permeate the gel and to exclusion of the electrolyte, it is usually limited in application to compounds of rather small molecular weight. As less dense resins are used which permit permeation of higher molecular weight substances, the ion exclusion becomes less efficient due to dilution of the fixed charges.

Although the process is theoretically indifferent to the counterion present in the resin, ion switching will occur if counterions other than those on the resin are present in the feed stock.

Many examples of compounds with high K_d values, due to affinity for the resin, are known.

2. *Product Recovery*

Many of the most exciting uses of ion exchange resins are in the recovery of products from natural sources. Frequently the desired compound is present in low concentration and often it is mixed with a great number of diverse compounds. The simple step of adsorption and elution under the grouping conditions outlined above results not only in a large reduction in the volume to be handled, but frequently in greatly improved purity of the product.

Group separations by ion exchange are readily scaled up and have been adapted to the production of chemicals on a tonnage basis.

Considerable experimental data are required for the design of an efficient production unit. These include optimization of flow rates and column size, regeneration and elution studies, and the effect of each variable on the subsequent steps of the process. However, for use in many laboratory isolation problems and in the use of an ion exchange step as a group separation procedure, satisfactory results may be obtained even though all variables are not optimized.

The general procedures for group separation as governed by pK and stability have been outlined above. The actual experimental approach may be influenced somewhat by availability and accuracy of assays and by knowledge of the desired compound.

Since one of the objectives in group separation is often reduction in volumes to be processed, an adsorption condition favoring high resin loading with minimum loss into the spent stream is usually desirable.

The overall selectivity coefficient has been discussed in Chapter II in terms of the dissociation constant of the material being adsorbed and considered in a pH range in which ionization of the resin is not affected. For basic substances, this is

$$\log K_{sel} = \log \frac{[RSO_3^- \ B^+]}{(BOH)} + \log \frac{(M^+)}{[RSO_3^- \ M^+]} - \log K_b - p(OH) \qquad (V.6)$$

in which $K_b = (B^+)(OH^-)/(BOH)$.

Although strict application of the above relation requires substitution of ion "activity" for all ionic materials within and without the resin, some insight into the interplay of pH and the dissociation constant of B can be obtained by assuming that "activity" and total content are equal (i.e., $f =$ ion/total $= 1$). If the absorption medium contains a high concentration of (M^+) so that the ratio $(M^+)/[RSO_3^- M^+]$ is not appreciably affected by the displacement of a low percentage of (M^+) from the resin (up to 20%), this may be considered constant and the logarithm is nearly 0. Therefore, if $K_{sel} = 1$, 99% of B would be adsorbed by the resin at a pH two units lower than the pK_b and 99% desorbed at a pH two units higher than the pK_b. At constant pH a tenfold increase in (M^+) would cause a tenfold decrease in the absorption. Thus a tenfold change in background electrolyte is equivalent to a change in pH of one unit. These approximations can be useful only around the pK value for the base and only if the pH is far away from the pK of ionized groups of the resin. In actual practice the simple case outlined above is rarely encountered. The actual activities of the ions are never equal to the total ionic strength. Frequently the ions involved are multivalent, leading to exponential factors in the equilibrium constants. In addition, there are usually several exchangeable components and each enters into the competition for resin sites. Finally, the resin loading may be substantially greater than 20%. This not only increases competitive pressure from displaced ions but also may enhance differences in resin site affinity due to an essential difference among the sites themselves.

A. Column Application. An example of high resin loading is the adsorption of glutamic acid from fermentation broth (*13*). Glutamic acid, $HOOC(CH_2)_2CHNH_2COOH$, has a molecular weight of 147 and an isoelectric pH of 3.2. The resin employed is Amberlite IR–120, a sulfonated polystyrene resin with 1.9-meq/ml exchange capacity. The reactions employed are as follows:

(1) *Adsorption or resin loading*

$$RSO_3^- \ldots H^+ + \underset{\substack{|\\ NH_3^+}}{RCHCOO^-} \rightarrow RSO_3^- \ldots NH_3^+ - \overset{R}{CHCOOH}$$

(2) *Elution*

$$RSO_3^- \ldots NH_3^+ \overset{R}{CHCOOH} + Na^+OH^- \rightarrow RSO_3^- \ldots Na^+ + \underset{\substack{|\\ NH_3^+}}{RCHCOO^-} + H_2O$$

(3) *Regeneration*

$$RSO_3^- \ldots Na^+ + H_2SO_4 \rightarrow RSO_3^- \ldots H^+ + Na_2{}^+SO_4{}^= + Na^+HSO_4{}^-$$

For the regeneration step resin containing 51 g equivalents of exchange capacity was treated with 55 g equivalents of 1.81 *N* sulfuric acid. The regenerated resin in a column 4 x 240 inches was treated with a solution containing 11 g equivalents of glutamic acid and 18.4 g equivalents of other cations. The spent liquor contained 0.06 g equivalents of glutamic acid (0.55%). After warming to 50°–60°C with a warm water wash, the column was eluted with 25.8 g equivalents of 0.9 *N* sodium hydroxide. The eluate contained 99% of the glutamic acid charged to the column. The pH at the end of the elution was about 7. Therefore, the eluate contained both glutamic acid and the sodium salt. Strong bases and most metal cations remain on the resin during the elution. The procedure concentrates weak bases in preference to strong bases. The resin loading was about 60 g of glutamic acid per liter.

In this single-column procedure the resin loading is not as high as might be achieved by continuing the feed until the breakthrough concentration of the glutamic acid is as high as the feed solution. It should be pointed out that the only selectivity involved is the H/Glu ratio and that selectivity with other ions is not involved since most of these are retained by the resin. The relatively high loading obtained in this circumstance would not be possible if the feed solution contained a much higher ratio of exchangeable cations to glutamic acid.

B. BATCH USE OF RESIN. Ion exchange resins are usually used in columns since the operations of adsorption and elution are easily carried out and increased efficiency is obtained by the countercurrent operation. However, in some cases batch or slurry operation may eliminate difficult filtration or concentration steps and give a satisfactory product.

Such a case is the recovery of heparin from hog casing mucosa (*14*). In this case a suspension of 4.5 kg of hog mucosa per liter of water containing 200 g/liter of ammonium chloride was coagulated by heat, diluted to about 8 liters and homogenized. The resulting homogenate contained about 550 g of mucosa and 27 g of NH_4Cl per liter. The heparin was adsorbed by adding 20 ml of Dowex 1 × 1 per liter of slurry and stirring for 16 hr. The resin was separated by passing the slurry through a wire screen which retains the resin but not the tissue slurry. After washing with water, the heparin was recovered in 73% overall yield by elution with 20% sodium chloride. After precipitation with methanol, which removes most of the sodium chloride, the weight of heparin recovered was about 440 mg/kg of hog mucosa. This corresponds to about 12 mg/ml of resin. Since heparin is a polysaccharide containing sulfate groups it is not possible to calculate the efficiency of the ion exchange reaction.

A major advantage of this procedure is the elimination of the difficult and tedious filtration step. In addition the product obtained is of high potency and requires minimal treatment to yield a final concentrate suitable for pharmaceutical use.

C. USE OF CARBOXYLIC ACID RESIN. Carboxylic acid resins are suitable for the recovery of amines having $pK_b > 9$. The resins were originally developed for the recovery of streptomycin but have since found application in the recovery of most basic antibiotics and in the separation of peptides. Although considerable variation in the degree of cross-linking and hence permeability to large molecules may be obtained, Amberlite IRC–50, produced by Rohm & Haas, readily takes up streptomycin, m. wt. 585, to a loading of 1 g/g of resin.

Multivalent basic antibiotics are readily recovered from dilute sources. Although streptomycin cannot readily be eluted from the resin by bases or high concentrations of salt, it is eluted by acidifying the resin. The eluate contains all the cations present in the resin adsorbate.

Other basic antibiotics, such as neomycin, m. wt. 800, are readily taken up. Since neomycin is a weak base, it can be recovered almost ash-free by elution with dilute ammonium hydroxide. This elution procedure may also be used with resin adsorbates made from strongly acidic resins. Frequently, however, carboxylic resins have higher capacity and better selectivity (i.e., lower losses into spent streams) than many sulfonic acid resins.

Many aminoglycoside antibiotics have amine groups with nearly identical pK values (about pK 9). For these antibiotics the resin selectivity is influenced greatly by the number of amino groups per molecule. Actinospectacin, which contains only two basic groups,

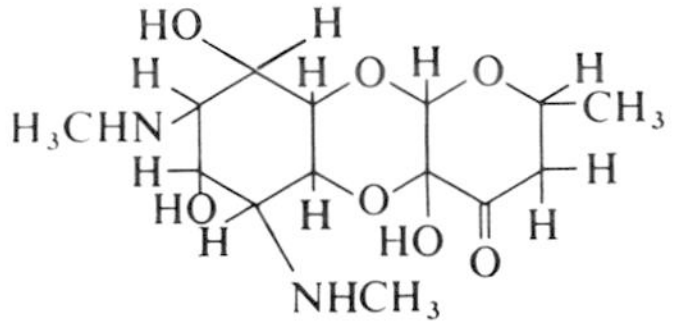

Actinospectacin

has been isolated using Amberlite IRC–50.

Although the molecular weight is not too high for ready permeation of the resin phase, the weakly basic properties of both the resin and the base prevent efficient adsorption. The selectivity obtained with this resin is very nearly the lower limit for applications in which the primary function of the resin is product recovery.

i. Effect of competing ions. In most applications to natural products a number of "competing" ions are usually present. The addition of minerals to fermentation broth may be necessary for optimum production yield and salts may be added during pH adjustment to facilitate extraction or filtration. Since loss of the product into the spent stream must be minimum and the concentration is usually low, a high "selectivity" toward the resin phase is required so that small resin volume can be used. The influence of divalent ions

TABLE V.3

ADSORPTION OF ACTINOSPECTACIN ON AMBERLITE IRC–50[a]

Mg^{++} (meq/ml of resin[b])	Ca^{++} (meq/ml of resin[b])	Actinospectacin (meq/ml of resin[b])	% Adsorbed
0	0	0.2	100
0.42	0	0.2	100
0.62	0	0.2	99.5
0.88	0	0.028	14
0.62	0.12	0.18	88
0.62	0.25	0.165	82.5
0.62	0.38	0.11	55

[a] pH 6.0, 25 ml resin/liter (Na-H form about 1 meq/ml).
[b] Assuming all added divalent ions are exchanged.

on actinospectacin adsorption is indicated in the Table V.3 (*15*). It is apparent that divalent ions are more readily adsorbed than actinospectacin.

If the Mg^{++} and Ca^{++} ions are removed or sequestered by adding F^- to a fermentation broth, only 67% recovery of actinospectacin at a charge level of only 0.15 meq/ml of resin is obtained. At this load level the resulting eluate does not have a great concentration factor over the broth and if the elution is carried out with acid, the eluate would have a high ash content.

One of the difficulties in the purification of actinospectacin with carboxylic acid resins is the low dissociation constant of the resin. At lower pH's the actinospectacin would be more highly ionized but the resin would be less ionized. Selectivity over competing cations would probably improve at a lower pH using a strongly ionized resin.

D. USE OF SPECIAL ELUTING SOLUTIONS. Compounds containing hydrophobic portions may be strongly adsorbed to resins with a polystyrene backbone. If the group is aromatic the adsorption is somewhat greater. The well-standardized Moore-Stein analytical procedures for amino acids which utilize sulfonic acid–polystyrene resins and buffers of varying strength and pH are indicative of this general rule. For simple α-amino acids with no other functional groups, the order of elution is glycine, alanine, α-aminoisobutyric acid, valine, methionine, isoleucine, leucine, tyrosine, phenylalanine, and tryptophan. Adsorption by the polymer backbone is illustrated by retention, in 0.3 *N* hydrochloric acid, of 5-hydroxymethyl-2-furaldehyde by sulfonic acid resin (H^+ form) and the even greater retention of 2-furaldehyde and salicylaldehyde in the same system (*16*).

For compounds with strong ionic attraction and additional adsorption, the high efficiency of adsorption and the possibility of stripping many less strongly adsorbed materials from the resin prior to elution of the desired substance indicate that recovery by ion exchange procedures would have operational advantages. However, such compounds may be difficult to recover from the resin adsorbate. High concentrations of salt, which might be difficult to separate from the product, are frequently of no value in eluting such materials.

One procedure, which has been used with some success, is the addition of water-miscible solvents to the elution solution. The optimum conditions vary somewhat with each substrate–resin combination but most substances can be recovered with aqueous solvent–salt mixtures.

i. Solvent–salt combinations. The use of solvent–salt solutions for the elution of tightly bound, relatively nonpolar organic compounds is detailed by

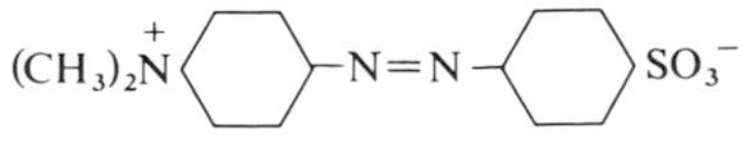

$$(CH_3)_2\overset{+}{N}\!-\!C_6H_4\!-\!N{=}N\!-\!C_6H_4\!-\!SO_3^-$$

Methyl orange

Wolf *et al.* (*8*). Methyl orange is particularly difficult to remove from anion exchange resins. However, 96% could be recovered from Amberlite IRA–411 (a strong base–polystyrene resin) with a mixture of 70% methanol, and 5% ammonium chloride. With Amberlite IRA–45, a weak base resin, a 68% recovery was obtained with 93% methanol, 5% ammonium chloride. Less than 0.1% is removed from the resin with four bed volumes of either aqueous NH_4Cl, 70% methanol, or 93% methanol. The influence of nonionic bonds is indicated by the lack of elution with 1 *N* sodium hydroxide, which depresses the ionization of the weak base resin.

ii. Organic salts. Using a Dowex 1 × 2 adsorbate of prednisolone phosphate less than 2% is eluted with aqueous 5% sodium chloride.

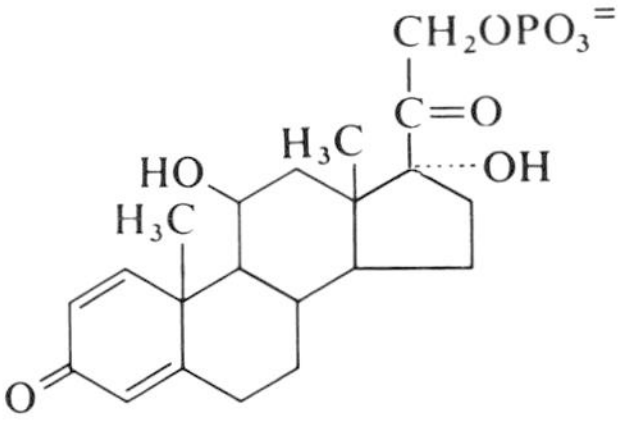

Prednisolone phosphate

Thus, many contaminants can be removed from the resin. The steroid is recovered in 83% yield with a mixture of 93% methanol and 5% ammonium chloride.

Many peptides are difficult to remove from ion exchange adsorbates. Mixtures of pyridine and acetic acid of varying molarity and pH have been found to elute almost all peptides (up to about 20 amino acid units) from such adsorbates. In extreme cases the eluant may be nearly 50% pyridine or acetic acid. These eluants can be removed by evaporation and have been adapted to gradient procedures for both anion and cation resins (*17*).

The importance of the co-ion has been mentioned. The removal of tightly bound ions from ion exchange resins or the selective elution of certain ions has been obtained by using eluants capable of sequestering the substance being eluted. Although this procedure has been successfully applied in the separation of metallic cations, the application to organic compounds is certainly possible.

E. METATHESIS REACTIONS. The application of ion exchange procedures to simple metathesis reactions may provide a quick and absolute process for converting ionic substances from one salt form to another without introducing undesirable substances into the solution being treated. The procedure constitutes one of the initial applications of ion exchange resins in the pharmaceutical industry (*18*).

Generally, the desired material does not take part in the exchange reaction. Thiamine mononitrate has more desirable stability properties than thiamine chloride or bromide, the product obtained by synthesis.

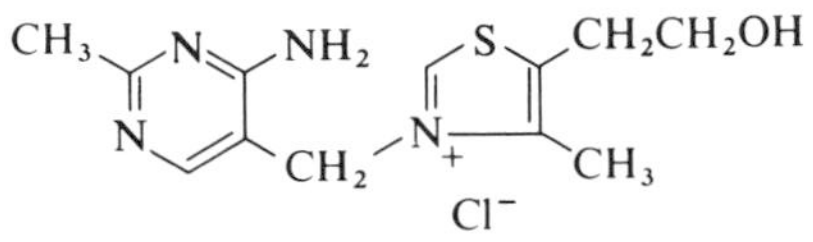

CH_3 N NH_2 S CH_2CH_2OH
N CH_2 N^+ CH_3
Cl^-

Thiamine chloride

Although thiamine chloride can be converted to the nitrate by treatment with silver nitrate and filtration of the precipitated silver chloride, any excess of silver nitrate added remains in the product. The same conversion is readily accomplished by using anion exchange resin in the nitrate form.

F. NEUTRALIZATION. Excess acid or base can be removed from process streams by ion exchange resins without adding additional acid or base to the solution. Kanamycin, a basic antibiotic containing three basic groups, is eluted from Amberlite IRC–50 by the addition of mineral acid. Kanamycin monosulfate, pH about 8.0, crystallizes on the addition of solvent whereas acidic salts are more difficult to isolate in crystalline form. The eluate on treatment with ion exchange resin in the hydroxide form can be brought to the pH needed for the crystallization without adding contaminants which may interfere with the crystallization. The resin can be used in a stirred batch type reaction, adding only the amount of resin needed to achieve the desired pH, or a column procedure can be used. In this case the final pH adjustment can be made by adding acid to the eluate.

Ion exchange reactions can also be used to prepare ash-free acid or free base forms of polar substances.

3. Fractionation Separations

Although group separations are probably of greater commercial significance than fractionation type separations using ion exchange, the latter are of much greater importance in the scientific investigation of natural products. As in solvent extraction, fractionation separation is required when the β, the separation factor of the materials being separated, is less than about 10.

The scale of the separation may be analytical (micrograms to milligrams) or preparative (milligrams to grams). In general any separation using ion exchange can be scaled up in size with no particular difficulty, and satisfactory operation obtained. Since diffusion in the solid phase is usually important in determining plate height in ion exchange processes, linear velocity or flow rate is an important consideration in scaling up.

The fundamental basis for fractionation separation with ion exchange is the distribution coefficient of the materials being separated between the liquid and solid phase. Usually the procedure is chromatographic with the resin in a column and the mobile phase passing through the column. As with other separation processes the separation factor or β required for the separation of two materials determines the effective number of stages or plates in the column. Since this may be diffusion-controlled, the effective number of plates in the column may be different for different substances.

As in other separation processes, nonlinear distribution coefficients produce skewed or tailed zones. Many of the factors contributing to nonlinearity have been mentioned previously and indeed complete linearity cannot be expected in any ion exchange process. However, with careful study conditions giving nearly linear distribution coefficients have been obtained with amino acids, which cannot only be separated but can be determined quantitatively, assuming a Gaussian distribution. For this separation, step gradients are commonly used and very low loading, in terms of ion exchange sites occupied, is employed.

Ion exchange separation of amino acids illustrates the application of the variables available. These include pH, ionic strength, temperature, and solvent. Stepwise solvent and temperatue changes or gradient buffer and programmed temperature changes have been used. Several procedures are satisfactory for the separation of the common amino acids. Usually two columns are used, one for basic amino acids and the other for acidic and neutral amino acids (*19*).

Although examination of the procedures used and the variables employed in the amino acid separation is instructive in designing an ion exchange separation procedure, the method cannot be recommended as a preparative procedure and in fact illustrates some of the major problems. The most serious are (1) loading limitations and (2) contamination with eluant buffer salts. In this procedure the load of any single amino acid rarely exceeds 1 mg and the band emerging from even the "short" column is 4–5 ml of 0.2–0.3 *N* buffer. Hence, 1 mg of "pure" product is contained in a solution with 70–100 mg of buffer salt. Satisfactory desalting of such a solution and recovery of the product is itself a major project.

Since the success of the separation depends on control of both ionic strength and pH, buffers are necessary. For proteins and other substances of high molecular weight, gel filtration or membrane separation such as dialysis or ultrafiltration can be used to separate the buffer salts. For lower molecular weight substances, the use of "volatile" buffers has provided a method of utilizing the procedure. A series of such buffers which permits application of both ionic and pH changes has been introduced for the ion exchange chromatography of peptides. The buffers are mixtures of pyridine, picolines, and *N*-ethylmorpholine with acetic or formic acid and can be used with both cation and anion exchange resins. Certain ammonium salts are also sufficiently volatile for use in ion exchange separations.

Some of the commonly employed buffers are listed in Table V.4.

TABLE V.4

COMPOSITION (ml/liter) OF VOLATILE BUFFERS

Reagents	pH of buffers				
	3.1	5.0	8.0	8.3	9.3
Pyridine	16.1	161.2	11.8	10	7.5
Acetic acid	278.5	143.2	0.1–0.2	0.4	0.1–0.5
α-Picoline	—	—	28.2	—	—
2,4,6-Collidine	—	—	—	10	—
N-Ethylmorpholine[a]	—	—	—	—	12.5

[a] N-Ethylmorpholine should be redistilled before use.

Other combinations of reagents may be used to achieve any pH within this range. Depending on the substance being purified ammonium salts may be satisfactory buffers. For example, ammonium chloride can be used to displace basic compounds from cation exchange resins. If the material is stable at pH 10.5–11, passage of the solution through an anion exchange resin in the hydroxyl form removes Cl^- and produces ammonium hydroxide which can be evaporated.

Buffer salts can sometimes by removed by a subsequent processing step without causing serious problems. Proteins may be precipitated by ammonium sulfate or solvents such as methanol or acetone. Solvent–water crystallizations may be carried out if the salt is sufficiently soluble in solvent employed. Adsorption by common adsorbents, such as carbon, alumina, or silica, may be satisfactory in the presence of salts.

When screening ion exchange methods or considering application in a particular problem, the problems of buffer removal should always be considered.

A. CHROMATOGRAPHIC PROCEDURE. The usual procedure for fractionation separation depends on careful selection of elution conditions. The adsorption generally does not contribute to the separation and is usually carried out under conditions which produce a narrow adsorbate band at the top of the column. For example, amino acids are usually applied to the column at pH 2–2.2, whereas elution is at pH 3.25 or higher.

Two elution procedures may be used. These are constant composition and gradient, either stepwise or continuous. As a general rule, the preparation of a single pure substance is best carried out by finding a combination of resin an eluant of constant composition which can be applied to the desired substance. The procedure allows higher loading and better fractionation than may be obtained by a gradient procedure. However, if each of several components is desired a gradient procedure may be required since some materials would appear too soon and others too late with a constant composition mixture. Gradients are sometimes used with substances having nonlinear distributions in order to improve yield and band sharpness. However, the required matching of the gradient and distribution coefficient may be difficult to achieve.

Although gradients are commonly used in protein fractionation the procedure has little to recommend it as a method of obtaining a single pure product. The gradient procedure may be used as a screening method to determine desorption conditions for the desired substance.

i. Buffer composition and distribution coefficient. A general procedure for the application of constant composition buffer in ion exchange chromatography consists in determining the approximate distribution coefficient by a series of batch experiments and then applying the selected condition in column chromatography. If the material can be detected on either thin-layer or paper chromatographs, thin-layer plates or paper loaded with ion exchange resin may be used.

A resin–buffer combination in which the distribution is about 10 to 1 in favor of the resin phase may be selected for the chromatography. Under these circumstances, a large number of effective transfers may be obtained without excessive dilution of the eluted band. In the batch equilibrium studies, a variety of resin types and eluants may be tried. A particular separation may be carried out successfully with one type of resin and not with another, depending on the factors influencing the distribution. If the compounds have only slight differences in pK, a resin having a potentially adsorptive polymer backbone, such as polystyrene, may be useful as the secondary adsorption effects may contribute substantially to the separation.

If it is assumed that the nonionized form of the material being studied does not adsorb on the resin and that the pH conditions do not influence the ionization of the competitive counterion and the resin, the mass action selectivity coefficient and the dissociation constant may be combined to determine the effect of pH and competing counterions on the proportion of material in the resin phase.

For the reaction

$$R^-A^+ + B^+ \rightleftharpoons R^-B^+ + A^+$$

$$K_{B/A} = \frac{[R^-\ B^+](A^+)}{[B^-\ A^+](B^+)}$$

and for substance B^+

$$K_{BOH} = \frac{(B^+)(OH^-)}{BOH}$$

$$p = \frac{\text{Amount in resin (vol resin} \cdot C_r)}{\text{Amount in resin + amount in solution (vol solution} \cdot C_s)}$$

Let

$$C_r/C_s = fK_{B/A}$$

at a constant ratio of solution volume to resin volume and assume that BOH does not adsorb on the resin, then

$$p = fK_{B/A} \frac{[R^-A^+]K_{BOH}}{(A^+)(OH^- + K_{BOH})} \tag{V.7}$$

If the resin is primarily in the $[R^-A^+]$ state (less than 20% substitution), then $[R^-A^+]$ and $fK_{B/A}$ can be considered constant (X_1) and the equation may be rearranged to yield

$$\frac{1}{p(A^+)} = \frac{1}{X_1}\left[\frac{(OH^-)}{K_{BOH}}\right] + \frac{1}{X_1} \tag{V.8}$$

A corresponding relationship for anion exchange in which K_{BOH} is replaced by K_{HA} and (OH^-) by (H^+) can be developed. Since K_{BOH} is also constant, the proportion of B in the resin phase increases as the pH is diminished and decreases as the concentration of competing ions is increased.

Equation (V.8) can be used for extrapolation using logarithmic plots of observed points. If the volume fraction is corrected to that existing in a column cross section (40% liquid, 60% resin), the value $1/p$ gives the number of displacement volumes of eluate expected.

The application of this general procedure for the determination of the dissociation constants of hydroxy acids using a system of anion exchange resin acetate is illustrated in Fig. V.5 (*20*). This figure also shows the dependence of the distribution coefficient on the concentration of (H^+) and (Ac^-).

Horvath *et al.* (*21*) has pointed out that the fractional retention of nucleotides, R, is approximately linear with buffer concentration for the nucleotides 3′-AMP and 3′-UMP and a specially prepared quaternary resin. Since retention, R, elution volume, V_e, and column void volume, V_0, are related by the relationship $V_e = V_0/R$, this provides a convenient method of extrapolation to a desired chromatographic elution volume. Some typical data are included in Fig. V.6.

Because of the assumptions and approximations inherent in the relationship as developed above, the distribution curves are usually not linear over wide ranges of pH or concentration and are, of course, influenced by the resin loading. However, as a method of determining suitable operating conditions for chromatography the procedure can be very useful.

In conducting exploratory batch equilibrium studies, it is necessary to

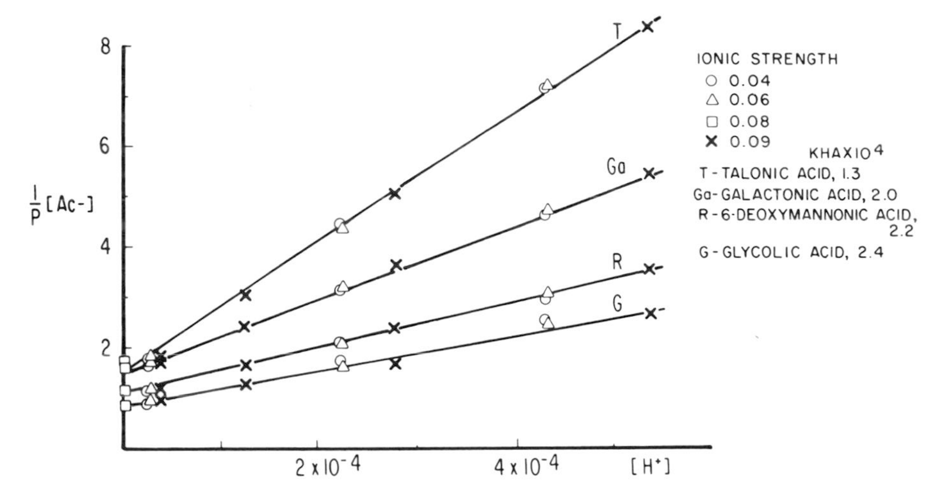

FIG. V.5. Effect of pH and electrolyte concentration on the distribution coefficient for hydroxy acids.

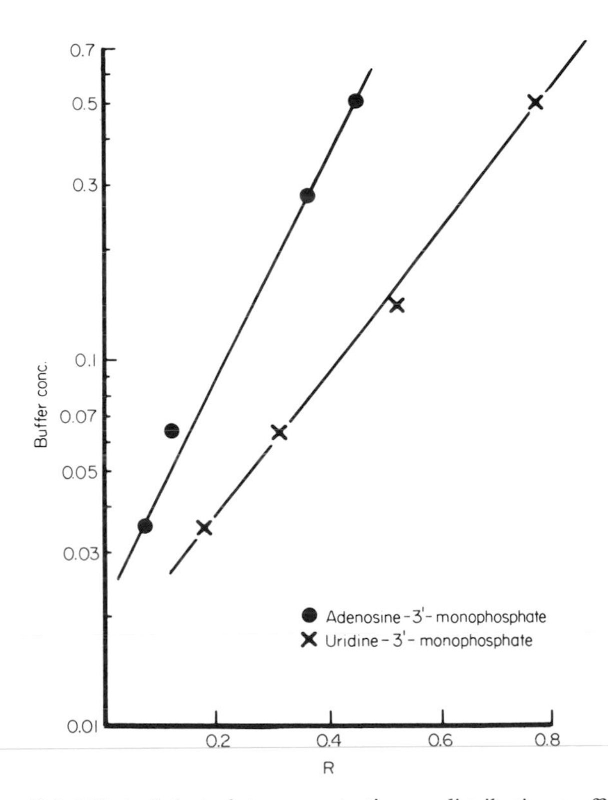

FIG. V.6. Effect of electrolyte concentration on distribution coefficient.

pre-equilibrate the resin with the buffer. Added materials should not exceed 10% of the total exchange capacity of the resin and if possible should be in the range of 2–5%. This condition limits the maximum concentration and total amount of added materials. If a buffer of low ionic strength is being used, the added substances should not cause a change in the buffer concentration of greater than twofold.

If the compounds being studied are nonpolar, high concentrations of electrolytes may tend to increase, rather than decrease, the distribution coefficient in favor of the resin phase. The addition of water-miscible organic solvents to the elution solution may give a satisfactory mixture.

Although for strongly ionized resins the pH of the eluant usually does not affect the ionization of the resin, weakly ionized resins, such as DEAE (diethylaminoethyl) cellulose and CM (carboxymethyl) cellulose are sometimes used at pH values at which the ionization of similar groups in free solution would be incomplete. Mechanical factors such as resin particle size and uniformity, column packing, and flow rate influence the resolution obtained. The volume changes exhibited by some resins with changes in buffer strength may cause loss of resolution by promoting channels in the resin bed. This change is more pronounced with resins of low equivalent weight and low cross-linking. The number of theoretical plates obtained in a particular system is influenced by the diffusion of the adsorbed substance in the solid phase. This has been investigated for cations by self-diffusion methods using radioactive tracers (*22*). These studies indicate, in general, that the ratio of the diffusion coefficient in the solid to the diffusion coefficient in the liquid is about 0.1. Values obtained by Hamilton *et al.* (*23*) for amino acids are in the range of 0.01. For this reason, linear velocity influences the efficiency of the separation. Choosing a distribution coefficient which requires passage of large volumes of eluting solution through the column results in increased operation time at the same linear flow rate or, if the flow rate is increased, causes increased band spread.

ii. Factors influencing column performance. The performance of ion exchange columns as a function of particle diameter, length, and linear flow rate has been investigated by Hamilton for the separation of amino acids. For a particular system, the distribution coefficient was shown to be independent of column length. In accordance with chromatography equations, the band spread [σ] for a particular substance was shown to increase as the $L^{\frac{1}{2}}$, $v^{\frac{1}{2}}$, and $d_p^{\frac{1}{2}}$ in which L is column length, v is linear flow velocity, and d_p is particle diameter.

After the distribution coefficient has been established and a particular buffer system selected for the chromatography, the variables of column length, linear velocity, and particle size can be manipulated to control the separation. For uniformly packed columns, the interdependence of the flow rate and column length indicate that with resin of a particular particle diameter, approximately the same degree of separation is obtained in the same time regardless of column length. If linear flow rate is decreased this has the same effect on resolution as lengthening the column while maintaining constant linear flow rate. Therefore the best procedure for increasing resolution in a fixed time is to decrease particle size. If a distribution coefficient is selected so that the material emerges from the column after about 25 void volumes and if the separation is complete in 24 hr, this is one void volume per hour. For a 100-cm column that is a linear flow rate of about 3×10^{-2} cm/sec. As a general rule the finest particle size resin available and operable should be used. Since the flow rate is influenced by the viscosity of the medium and the compressibility of the resin, it may be desirable to adjust the distribution to a value readily obtained in the available equipment.

The fraction size collected from an ion exchange chromatography may be an important operating variable. If too few fractions are collected the resolution may be lost because of mixing after the column. On the other hand, too many fractions may cause serious analytical problems. As a general rule, about 10 fractions should be collected for each column void. The zone containing the desired compound can be detected by analysing every three to five fractions and a more accurate determination of the actual band width made after the zone is located initially.

If possible the effluent stream should be monitored by some sensitive detection procedure. Various physical properties such as ultraviolet absorption, refractive index, and conductivity have been used, depending on the properties of the material being fractionated.

4. Applications of Ion Exchange Chromatography

A. CONSTANT COMPOSITION ELUANT. Although there are many examples of the use of ion exchange chromatography in the separation of rare earths, examples with organic compounds other than proteins or peptides are limited. One of the earliest applications was in the separation of purine and pyrimidine bases. Cohn (*24*) showed that the common bases, uracil, cytosine, guanine, and adenine, can be separated by ion exchange chromatography using either cation exchange or anion exchange resins.

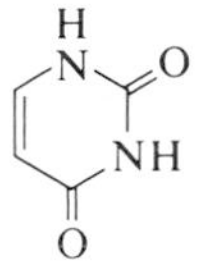

Uracil	Cytosine	Guanine	Adenine
p*K*=9.45	p*K*= 4.50	p*K*= 1.9	p*K*=4.1
	12.2	9.0	9.8
		12.2	

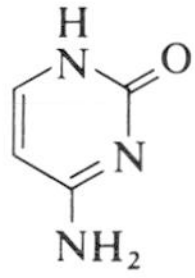

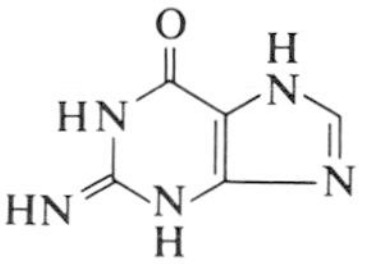

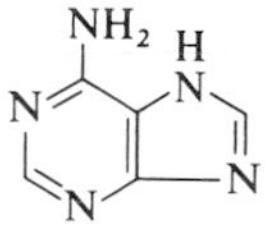

Although many procedures have been developed and the orginal methods refined, the procedures used originally are indicative of the great flexibility of ion exchange chromatography. Since the compounds have high ultraviolet absorption very dilute solutions can be readily detected. The chromatography was carried out using Dowex 50 resin in the H^+ form and the column was developed with 2 *N* hydrochloric acid. A short column, 8.1 cm, was used at flow rate of 3×10^{-2} cm/sec. Resin of small particle size, about 50 μdiameter, was used. However, the distribution coefficients were much larger than recommended above. The bases are eluted in the order given above. Elution volumes of about 8, 18, 30, and 80 bed volumes were obtained.

At an elution rate of 1 bed volume in 10 min (a consequence of the short column used) the time required for the chromatography is about 15–16 hr. The last material eluted, at about 80 bed volumes, is contained in a volume of almost 30 bed volumes. Under this circumstance a step change in the elution would be desirable to improve the dilution of this component. Since the displacing substance is H^+ and is already present at 2 *N* concentration there is no simple procedure for improving this system.

However, the resin can be converted to the Na form and the chromatography carried out with a buffer at higher pH, which would diminish the K_d for all the bases. With citrate buffer at pH 4, 0.2 *N* (*24a*) with a gradient to 1 *N*, the same bases were eluted at about 1, 4.5, 7, and 8.5 bed volumes. Guanine appeared before cytosine in this case.

The same four compounds were also separated on an anion exchange resin. In this case Dowex 1 resin in the chloride form was used. The elution order with 0.2 *M* NH_4OH, 0.025 *M* NH_4Cl buffer at pH 10.6 is cytosine, uracil, guanine, and adenine. After guanine is eluted, the adenine is obtained by increasing NH_4Cl to 1 *M*. The products were obtained in about 4, 25, 75, and 100 bed volumes, respectively.

In these examples very low loadings of the materials being separated were used and only 0.1–0.5% of the total exchange capacity of the resin was utilized. Although separation on this scale may be useful as an analytical method, recovery of pure product from the large volumes of eluant would be difficult on a preparative scale. However, the procedure can be readily adapted to larger scale.

If an arbitrary *p* value of 0.1 were selected for the desired substance, cation exchange at low pH could not be used for any of the compounds. However, by using cation exchange at higher pH conditions producing $p = 0.1$ for any of the bases can be obtained. Under this condition a much higher loading of the resin can readily be employed. Although the K_d is probably not linear with resin loading, it is reasonable to assume that nonlinearity would result in decreased affinity for all the compounds. If a total charge of bases as high as 25% of the total capacity of the resin is used (with a buffer–cation exchange resin system), it is instructive to consider the probable distribution of compounds on the column at the end of the loading cycle. If the distribution were chosen for obtaining uracil, the pH would be rather low and all of the unwanted compounds would be located behind uracil. Because of their greater affinity for the resin the uracil zone would not be seriously contaminated with the other materials and development of the column would continuously move the desired compound ahead of the others and good product would be obtained. On the other hand if conditions were chosen such that adenine had a p value of 0.1, the other materials would precede adenine and again good purity would be obtained, even at high loading. However, if much more cytosine than adenine were present, it would contaminate the adenine seriously on cation exchange resins but would be readily eliminated by anion exchange. If guanine were the desired substance, the loading and chromatography would need to be carefully controlled since it is midway between uracil and cytosine. However, it would be displaced by large quantities of adenine and would in turn displace uracil. Thus, the method can be scaled up readily if only one of the compounds is sought.

B. GRADIENT ELUTION PROCEDURES. Gradient elution procedures are used much more commonly than constant composition or stepwise procedures. The method is used in protein, peptide, and nucleic acid separations. Although the procedure overcomes dilution due to low K_d and may prevent tailing, the necessity for choosing the proper gradient for the separation being carried out frequently requires excessive laboratory time to achieve a good separation. The procedure has the virtues that usually the compound is recovered in good yield and that some purification is usually obtained.

Gradients are usually generated by adding one or more components to a mixing vessel containing the starting solution. Depending upon the geometry of the vessels employed and upon whether the mixing vessel contains a constant volume, all types of gradients, linear with variable slope and concave or convex, can be produced. The steepness of the gradient must be adjusted to the column. If the gradient is too shallow the desired prevention of tailing may not be obtained and if it is too steep the separation may be incomplete.

Gradients of ionic strength may cause shrinkage of the resin bed with a resulting decrease in resolution.

If the gradient employed is pH, the ion exchange resin may act as a buffer and effectively diminish the gradient until a breakthrough point is reached. When this occurs many adsorbed substances may unload at nearly the same time. Such an occurrence may produce a deceptive elution pattern since many of the characteristics of a Gaussian elution band are related to the mechanical factors contributing to zone spreading, and since a breakthrough of the type mentioned above merely reflects homogenity of the resin, a gross mixture of substances can appear as a homogeneous Gaussian band.

Methods of establishing the gradient needed for a particular separation are usually empirical. However, operational details may influence the experimental design. Assuming that the beginning and end of the gradient are known, the steepness can be arranged so that a specific volume will be used for the fractionation.

Pyridine–acetic acid gradients (*17*) have been useful in the separation of peptides containing up to 20 amino acids using Dowex 50WX2 resin. The starting buffer is about pH 3.1 and the final buffer about pH 5.0 (see Table V.4). Elution is due to charge reduction of the adsorbed peptide and the solvent effect of added pyridine. The procedure was applied by Schroeder for the separation of tryptic peptides obtained from hemoglobin. The column was prepared with Dowex 50X2 resin (200–400 mesh) which had been conditioned by successive washing with NaOH, HCl, pyridine, and finally pH 3.1 buffer. About 0.4–1 g of peptide mixture was chromatographed using a 3.5 × 100 cm column and a three-chambered gradient device. The mixing chamber contained 4.3 liters of pH 3.1 buffer and was connected to two vessels of equal size. The column temperature was regulated at 38°C and the flow rate maintained at 2 ml/min. About 10 liters of buffer divided into 1000 fractions was used in the run. Under these conditions the flow rate was about 8×10^{-3} cm/sec; 10 column volumes were used to carry out the separation, which required about 80 hr. In one run 22 peptide peaks were observed; some were not pure compounds.

A similar procedure using anion exchange resin (Dowex 1X2) and a gradient from pH 8.3 to pH 2.2 using a final solution of 2 *N* acetic acid has also been used with peptides.

For the separation of materials of higher molecular weight, resins based on celluose or cross-linked dextran have found wide acceptance. Since proteins are usually unstable at extremes of pH, operation is usually in the pH range of 4–8. The most common exchange groups are DEAE (diethylaminoethyl) and CM (carboxymethyl) (*25*). Although strongly basic groups have been added to cellulose, the titration behavior of these exchanges suggests that some modification of base strength occurs during reaction.

A variety of acid groups have been introduced, ranging from strongly acidic (sulfoethyl), through intermediate phosphates, to weakly acidic carboxyl.

For ion exchange celluloses the degree of substitution can be controlled. Reduction of the degree of substitution has been reported to improve the rate of adsorption and allow higher yields (*26*). Since the materials are usually fibrous (even if a spherical particle has been formed), the exchange sites appear less uniform than with synthetic polymer resins. It is with these exchange materials that gradient elution procedures have become standard.

The flow properties of ion exchange celluloses and of dextran ion exchange substances are usually poor since the materials are rather soft and deform under pressure. Careful packing of the columns is required and rapid changes of ionic strength which shrink or expand the resin network should be avoided.

c. PROTEINS. Proteins are amphoteric substances containing both cationic and anionic groups. Hence either anion or cation exchange resins may be used. Most proteins have isoelectric points within the pH range 4–8. At this point, the molecule contains as many anionic as cationic charges and therefore does not adsorb well to ion exchange resins. Since the molecule becomes cationic at pH values lower than the isoelectric pH and anionic at pH values higher than the isoelectric pH, a general rule of selecting an operating pH about two units from the pI may be used. If the isoelectric pH is 5, a cation exchange resin would be used at pH 3 or an anion exchange resin at pH 7. This pH requirement may limit the resin choice. In an initial experiment a gradient of ionic strength may be operated using the pH indicated. The pH may then be adjusted to improve the chromatography. Purification factors of 5–20 are usually obtained.

Pure proteins are usually obtained by a multiple-step fractionation procedure. Ion exchange chromatography is frequently required but recovery from the buffer salts may be tedious. The stability of proteins is usually influenced by both pH and temperature. As a general rule some relationship between pH and temperature exists and the point of maximum stability is near the isoelectric point. Hence ion exchange separations are operated at some pH different from that of maximum stability. For this reason such separations are usually carried out in the cold.

Many examples of the separation of different enzyme activities from a protein concentrate are contained in the literature. Frequently such separations do not yield any of the enzymes in pure form. In seeking a single pure substance, preliminary studies designed to determine the influence of pH on stability and the isoelectric point of the desired protein may be useful in designing an efficient chromatography.

L-Asparaginase, an enzyme which specifically hydrolyzes L-asparagine, has been shown to be effective in certain types of human cancer. The enzyme is produced by *E. coli*. Initial studies showed that the L-asparaginase activity contained in ammonium sulfate-precipitated protein from cell extracts derives from at least two proteins. One of the enzymes is inactive in cancer treatment and the other active.

The two enzymes were separated by chromatography on DEAE cellulose (*27*). In this procedure DEAE cellulose was equilibrated with pH 8 phosphate buffer. The column was developed with a linear gradient of NaCl. One enzyme appeared in the eluate at a point corresponding to 0.06 *M* sodium chloride and the second at a point corresponding to 0.2 *M* sodium chloride. In this case the separation of the two enzyme types was an important objective since it allowed the characteristics of each to be studied.

D. MACROMOLECULES. *i. Liquid ion exchange.* Ion exchange resins are not satisfactory with very large molecules, especially nucleic acids. Recently it has been shown that non-cross-linked electrolytes are useful in the separation of these compounds. The system of Kelmers (*28*) represents a hybrid between ion exchange and partition. The column packing consists of diatomaceous earth rendered hydrophobic by treatment with dimethyldichlorosilane. Two parts of the solid are mixed with one part of a 4% solution of dimethyldilaurylammonium chloride in isoamyl acetate. The mobile phase is 0.05 *M* tris buffer (pH 7.4) containing 0.3 *M* sodium chloride. The column is used with a linear sodium chloride gradient for the separation of t-RNA. The distribution coefficient of the RNA was shown to be influenced by the temperature, the pH, and the type of counterion. The addition of a low concentration of Mg^{++} to the mobile phase lowers the distribution coefficient.

ii. Modified cellulose or silica. A recently published procedure for the separation of the same t-RNA substances utilizes a new type of modified cellulose ion exchange resin. DEAE cellulose is benzoylated (*29*) and either used directly in solvent-containing mixtures or dissolved in methylene chloride and dispersed on silicic acid (weight ratio, 1–25) to improve availability of the exchange sites to the high molecular weight substance (*30*).

iii. Insoluble linear polymers. Native DNA has been fractionated using a column containing kieselguhr and polylysine (m. wt. 50,000) (*31*). The DNA was eluted at pH 6.7 using a gradient of sodium chloride (0.4 to 4.0 *M*). Peaks observed by ultraviolet absorption were shown to have a different composition of purine and pyrimidine bases. Although it is not known whether the separation is a result of precipitation of an insoluble polylysine complex and slow dissolution as the ionic strength is increased or whether the polylysine remains fixed to the support, this general procedure appears useful in the fractionation of macromolecules for which no procedure based on chemical properties is now known.

B. Separation of Neutral Compounds with Ion Exchange Resins

Ion exchange resins are useful in the chromatographic separation of neutral organic compounds. Two general procedures have been applied.

1. Formation of Ionized Complex

In the first, the separation is based on the formation of an ionizable complex. For example, sugars and polyols form acidic complexes with borate. By using anion exchange resins in the borate form and developing the chromatogram with borate buffer, a number of carbohydrates were separated (*32*).

Compounds containing carbonyl groups capable of reacting with bisulfite form complexes with anion exchange resins in the bisulfite form. With a quaternary resin aldehydes and ketones are adsorbed at room temperature. Ketones are eluted with water at 75°C. The more stable aldehydes are eluted with 1 *N* sodium chloride. The separation of lactaldehyde, acetol, and pyruvaldehyde by elution with potassium bisulfite of increasing molarity and recovery from the eluate has been reported (*33*, *34*). Vitamin B_{12}, a neutral compound containing cobalt, forms an ionized compound with alkaline solutions containing cyanide ion.

$$\text{Vitamin } B_{12} + 2\ CN^- \rightarrow [\text{Vitamin } B_{12}(CN)_2]^=$$

The complex binds with anion exchange resins and then can be separated from acidic materials which do not require cyanide for acidity and from neutral substances incapable of forming acidic cyanide complexes (*35*).

Many other methods of forming ionic complexes from otherwise neutral compounds are possible. The recent introduction of ion exchange resins (Amberlyst series) which maintain porosity in organic solvents permits the use of ion exchange reactions under anhydrous conditions.

2. *Retention of Nonionized Compounds*

A second procedure which has found extensive use with ion exchange resins is the adsorption or preferential extraction of neutral compounds by the polymer phase. This may be carried out using unmodified ion exchange resin in water, in water-miscible solvents, or in aqueous solutions containing large amounts of salt. The "solvent" power of the ion exchange resin may be modified by conversion to a less polar form, such as a fatty acid derivative.

In the orginal report of ion exclusion, Wheaton and Bauman (*12*) reported that the distribution coefficients of a number of neutral compounds were different and appeared to be influenced by the salt form of the resin. For example, the K_d $[K_d = (V_e - V_0)/V_i]$ of acetone was observed to be 0.66, 1.08, and 1.20 with Dowex 1X8 sulfate, Dowex 1X7.5 chloride, and Dowex 50X8 hydrogen, respectively. With Dowex 50X8 hydrogen, phenol had a K_d value of 3.08 whereas with Dowex 1X7.5 chloride a value of 17.7 was obtained. These preliminary data indicated that ion exchange resins were capable of partitioning neutral compounds. The differences in K_d can be amplified by modification of the system.

Partition between the solvent inside the resin and that in the mobile phase may contribute substantially to the separation obtained. In most cases the composition of the solvent in the two phases is substantially different. The distribution of water between the resin and solvent has been studied with a variety of resins and solvents (*36*). In general, the greater the solvent concentration employed the greater the difference. For example, with a polystyrene resin in the sodium form an equilibrium was attained with dioxane in which the external solution contained 30% water whereas the internal solution contained 83% water. This procedure was used by Samuelson and Swenson (*37*) to adjust the distribution coefficient to obtain the distribution required. Thus with 72% ethanol and Dowex 1X8 sulfate incomplete separation of glucose and lactose was obtained when 10 mg of each was charged to a column (1 × 84 cm). Separation was complete using 74% ethanol. The effect of solvent concentration on the distribution coefficient for a variety of sugars is given in Table V.5.

TABLE V.5

DISTRIBUTION COEFFICIENT CALCULATED FROM PEAK ELUTION VOLUME[a]

Sugar	Distribution coefficient[b] at % ethanol				
	65%	70%	72%	74%	82%
Xylose	—	—	—	8.17	—
Glucose	6.35	8.8	11.1	12.6	36.9
Cellobiose	—	12.1	—	19.3	—
Lactose	—	—	15.8	18.3	74.4
Maltose	—	—	—	18.3	—
Sucrose	—	—	14.8	17.9	—
Mellezitose	—	—	—	—	25.8
Raffinose	13.1	—	—	—	42.6
Stachyose	18.8	—	—	—	—
Verbascose	32.4	—	—	—	—

[a] On Dowex 1X8 sulfate.

[b] $K_d = (V_e - V_o)/V_r$, in which V_e is peak elution volume; V_o is column void; and V_r is resin volume. V_r/V_o is assumed to be 1.5.

The distribution coefficient is also influenced by counterions, the polymer backbone, and temperature. Cation exchange resins can be used with higher concentrations of alcohol (*38*). Compounds which would prefer the solvent phase of the above system can be "driven" into the resin phase preferentially by using high concentrations of a salt in the mobile phase. Since the salt is excluded from the resin by the Donnan membrane potential, compounds less soluble in the concentrated salt solution are retained by the resin. This phenomenon was termed "salting-out" and was used by Sargent and Rieman (*39*) to separate simple water-soluble alcohols by chromatography.

Organic acids and bases can be chromatographed on ion exchange resins under conditions in which ionization is not important. This was first observed with the basic antibiotic kanamycin, which was separated into three components by chromatography of the free base on Dowex 1X2 hydroxide form

(*6*). The degree of the separation is indicated in Fig. II.2. In this case the chromatogram is eluted with water. Buffer salts are not used and would probably detract from the separation since the antibiotic is not retarded by the resin in salt form. This procedure has been found to separate all of the basic antibiotics which occur as mixtures (*40*).

Using a sulfonated polystyrene in the hydrogen form a number of organic acids were separated in the system acetone–methylene chloride–water (20:15:1). With this system the less polar acids are eluted first (*41*). If water alone is used organic acids are retarded on sulfonic acid resin but the least polar are retarded most (*42*).

The relatively few examples of the application of ion exchange resins included above have been selected to show the broad spectrum of usage of these materials, the areas of most potential, and some of the procedures which have been used.

REFERENCES

1. Hale, D. K., Packham, D. I., and Pepper, K. W., *J. Chem. Soc.* pp. 844–851 (1953).
2. Kanhere, S. S., Shah, R. S., and Bafna, S. L., *J. Pharm. Sci.* **57**, 342–345 (1968).
3. Myers, G. E., and Boyd, G. E., *J. Phys. Chem.* **60**, 521–529 (1956).
4. Helfferich, F., "Ion Exchange." McGraw-Hill, New York, 1962.
5. Chaiet, L., U.S. Patent No. 3,163,637, (1964).
6. Rothrock, J. W., Goegelman, R. T., and Wolf, F. J., *Antibiot. Ann.* p. 796 (1958–1959).
7. Sargent, R. N., and Graham, D. L., *Ind. Eng. Chem. Process Design Develop.* **1**, 56–63 (1962).
8. Wolf, F. J., Putter, I., Downing, G. V., Jr., and Gillin, J., U.S. Patent No. 3,221,008, (1965).
9. Bergren, A., and Bjorling, C. O., *Acta Chem. Scand.* **11**, 179–180 (1957).
10. Tsuk, A. G., and Gregor, H. P., *J. Am. Chem. Soc.* **87**, 5538–5542 (1965).
11. Hatch, M. J., Dillon, J. A., and Smith, H. B., *Ind. Eng. Chem.* **49**, 1812–1819 (1957).
12. Wheaton, R. M., and Bauman, W. C., *Ann. N.Y. Acad. Sci.* **57**, 159–176 (1953).
13. Conklin, D. A., and Gillin, J., U.S. Patent No. 3,325,539, (1967).
14. Williams, R. E., U.S. Patent No. 3,337,409, (1967).
15. Jahnke, H. K., U.S. Patent No. 3,206,360, (1965).
16. Barker, S. A., Murray, K., and Stacey, M., *Nature* **186**, 469–470 (1960).
17. Schroeder, W. A., Jones, R. T., Cormick, J., and McCalla, K., *Anal. Chem.* **34**, 1570–1575 (1962).
18. Howe, E. E., and Tishler, M., U.S. Patent No. 2,597,329, (1952).
19. Moore, S., Spackmann, D. H., and Stein, W. H., *Anal. Chem.* **30**, 1190–1206 (1958).
20. Samuelson, O., and Thede, L., *J. Chromatog.* **30**, 556–565 (1967).
21. Horvath, C. G., Preiss, B. A., and Lipsky, S. R., *Anal. Chem.* **39**, 1422–1428 (1967).
22. Soldano, B. A., and Boyd, G. E., *J. Am. Chem. Soc.* **75**, 6107(1953).
23. Hamilton, P. B., Bogue, D. C., and Anderson, R. A., *Anal. Chem.* **32**, 1782–1792 (1960).
24. Cohn, W. E., *Science* **109**, 377–378 (1949).
24a. Crampton, C. F., Frankel, F. R., Benson, A. M., and Wade, A., *Anal. Biochem.* **1**, 249–262 (1960).
25. Peterson, A. E., and Sober, H. A., *Methods Enzymol.* **6**, 3–37 (1962).

26. Knight, C. S., Weaver, V. C., and Brook, B. N., *Nature* **200**, 245–247 (1963).
27. Campbell, H. A., Mashburn, L. T., Boyse, E. A., and Old, L. J., *Biochemistry* **6**, 721–730 (1967).
28. Kelmers, A. D., *J. Biol. Chem.* **241**, 354–355 (1966).
29. Gillam, I., Millward, S., Blew, D., von Tigerstrom, M., Wimmer, E., and Tener, G. M., *Biochemistry* **6**, 3043–3056 (1967).
30. Wimmer, E., Maxwell, I. H., and Tener, G. M., *Biochemistry* **7**, 2623–2628 (1968).
31. Ayad, S. R., and Blamire, J., *Biochem. Biophys. Res. Commun.* **30**, 207–211 (1968).
32. Khym, J. X., and Zill, L. P., *J. Am. Chem. Soc.* **73**, 2399–2400 (1951).
33. Huff, E., *Anal. Chem.* **31**, 1626–1629 (1959).
34. Christofferson, K., *Anal. Chim. Acta* **33**, 303–310 (1965).
35. Shafer, H. M., and Holland, A. J., U.S. Patent No. 2,709,669, (1955).
36. Davies, C. W., and Owen, B. D. R., *J. Chem. Soc.* pp. 1676–1680 (1956).
37. Samuelson, O., and Swenson, B., *Anal. Chim. Acta* **28**, 426–432 (1963).
38. Rückert, H., and Samuelson, O., *Acta Chem. Scand.* **11**, 315–323 (1957).
39. Sargent, R., and Rieman, W., III, *J. Phys. Chem.* **61**, 354–358 (1957).
40. Maehr, H., and Schafner, C. P., *J. Chromatog.* **30**, 572–578 (1967).
41. Seki, T., *J. Chromatog.* **22**, 498–499 (1966).
42. Miller, T. W., private communication, 1968.

GENERAL REFERENCES

"Dowex: Ion Exchange." Dow Chem. Co., Midland, Michigan, 1964.
Helfferich, F., "Ion Exchange." McGraw-Hill, New York, 1962.
Kunin, R., "Ion Exchange Resins." Wiley, New York, 1966.
Marinsky, J. A., ed., "Ion Exchange." Dekker, New York, 1966.

VI *Adsorption Processes*

I. GENERAL PRINCIPLES

A. Introduction

This chapter is concerned primarily with the selection and use of adsorbents in which the adsorption mechanism is basically a surface phenomenon. Adsorbents most useful in this process are activated carbon; oxides of silicon, aluminum, and titanium; and various types of adsorbent clays. Although no completely satisfactory unifying mechanism has been developed and exceptions can be found to almost any generality, this adsorption will be considered in terms of a monolayer of adsorbed material on the surface of the adsorbent. While recognizing that the surface of any adsorbent cannot be adequately defined either physically in terms of pores, mountains, and valleys or chemically in terms of the actual composition of the surface, the concept is useful in considering applications of solid adsorbents. The following discussion will consider the consequences of this hypothesis.

In general, the compounds being considered are either not volatile or not stable at the temperature required for volatility. Therefore only liquid systems will be discussed. In such systems the affinity of an adsorbent for the compounds being separated is dependent on the solvent employed.

The systems are limited, generally, to molecules of relatively small size (less than 2000 m. wt.) since much of the surface area of adsorbents is contained in pores which may be inaccessible to larger molecules, or accessible only at slow rates. Desorption of large molecules is usually difficult, due, at least in part, to multiplicity of interaction sites.

B. Adsorption Isotherms

For any particular system of adsorbent, solute, and solvent, the adsorbent surface is always covered. The relative affinity of the surface for the solute and solvent determines adsorption efficiency. Although adsorption isotherms represent equilibria, these are dynamic in that the equilibrium represents the point at which the rate of escape from the surface is equal to the rate of adsorption. Since no adsorbent surface is completely uniform either physically or chemically, certain portions of the surface attract solute or solvent with greater affinity than other portions of the surface.

A plot of the average concentration of the adsorbed solute and the concentration in solution (for liquid systems) is called an adsorption isotherm and will generally have the form illustrated in Fig. VI.1. The solid line represents

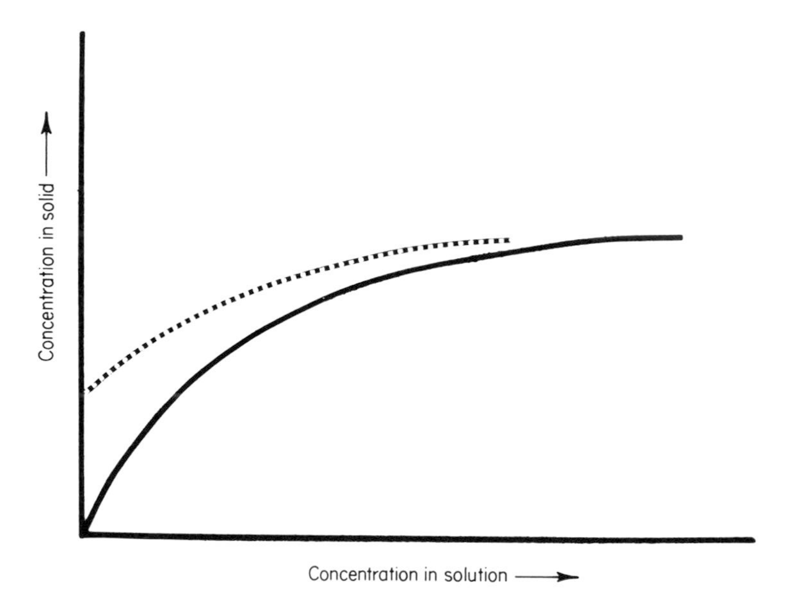

FIG. VI.1. General adsorption isotherm.

that usually observed. However, certain systems have been observed which appear to fit more nearly the dotted line. In such systems a portion of the surface may be very active in removing the solute although other explanations may be offered in specific cases. The occurrence of this type of adsorption with a natural product usually indicates that the system cannot be used for the recovery of the adsorbed material.

1. Langmuir Isotherm Equation

The most useful relationship explaining the shape of the adsorption isotherm was derived by Langmuir (*1*). The equation is based on the following assumptions:

(1) Monolayer adsorption

(2) All adsorption sites equivalent
(3) No interaction between adjacent adsorbed molecules

For application in liquid systems it is also assumed that all adsorption sites are occupied by either solute or solvent molecules.

Designating solute molecules, A, solvent molecules, S, and surface sites, Y, the adsorption process can be written as

$$A + SY \rightleftharpoons AY + S$$

and

$$K_a = \frac{[AY][S]}{[A]]SY]} \qquad (VI.1)$$

in which the quantities are expressed in terms of thermodynamic activities. If the sites are of equal activity, then the ratio AY/SY is the same as the ratio of the actual concentration of AY and SY in the adsorbent. If θ is the fraction of surface covered by A and $1 - \theta$ the fraction covered by S,

$$\frac{[AY]}{[SY]} = \frac{\theta}{1-\theta}$$

Further, since the concentration of A is normally low, the mole fraction of S in the liquid phase, $[1 - A]$, is approximately 1. Hence,

$$K = \frac{\theta[1 - A]}{[1-\theta][A]} \cong \frac{\theta}{[1-\theta][A]} \qquad (VI.2)$$

At low concentrations of A and small values of θ, the term θA is small and the ratio θ/A becomes nearly constant. At high concentrations of A in the liquid phase, θ approaches 1 and the concentration of A in the adsorbent becomes constant.

2. *Freundlich Isotherm*

Since the assumption concerning surface activity is usually not exact, many adsorption isotherms do not fit the Langmuir expression and other more complex relationships have been derived. A relatively common expression, the Freundlich isotherm, has the form

$$\theta = K_f C^n \qquad (VI.3)$$

in which n (< 1) and K_f are constants describing a particular system. In ordinary use the term θ is replaced by x/m, grams of solute per gram of adsorbent.

Other expressions for the adsorption isotherm have been shown to be applicable in certain systems and are sometimes necessary to account for the inhomogeneous nature of many adsorbents. Generally for group separation processes using the adsorbent in a batch method, the Freundlich expression is adequate for the evaluation of various adsorbents and solvents and for yield prediction under known operating conditions. If the separation is a fractionation type, the selection of conditions conforming more nearly to the Langmuir expression is desirable.

C. Factors Affecting Adsorption

The efficiency of the adsorption of a particular substance is influenced by the nature of the surface, the nature of the material being adsorbed, and the solvent used for the adsorption.

1. Surface Availability

If the surface area is completely uniform and available to molecules of any size, the percentage of the surface covered would be influenced by the shape, but not appreciably by the size, of the adsorbed molecule. That the surface area of activated carbon is either not uniform or completely available to molecules of different size is indicated by the fact that the surface area determined by total capacity decreases as the size of the solute increases.

Since the adsorbent must have a sufficiently large gross particle size to be filtered or packed in columns, a large portion of the surface consists of a pore network. The size and uniformity of these pores influences the adsorption of large molecules. It is therefore not surprising that the available area diminishes as the molecular size of the adsorbed solute increases.

2. Solute Size

For organic molecules containing several atoms the adsorption affinity increases as the number of sites participating in the adsorption of a single molecule increases. Thus, although the total area available for large molecules decreases, the actual affinity for large molecules may increase. With homologous series of fatty acids and activated carbon, this property forms the basis of several analytical systems.

3. Surface Modification and Structure

The surface area of most useful adsorbents is heterogeneous, both physically and chemically. This, in effect, means that the K of the Langmuir isotherm equation is not constant for a particular system. The surface heterogeneity can give rise to specific adsorption effects which may be quite useful. Frequently, however, nonuniform surfaces are a disadvantage since the conditions of use are governed by the most active sites and effective use of the remainder of the surface may be impossible.

Although a portion of the surface heterogeneity can be due to steric or geometric effects, the chemical nature of the surface is also important. Snyder (*2*) has studied the surface heterogeneity of silica and alumina as it affects the distribution coefficient of a solute in a particular system. In these studies the distribution coefficient was determined by the volume required to elute 50% of the applied material for a column. As the same substance is applied in increasing quantities the elution volume decreases as the area occupied by the solute decreases in affinity for the solute. The effect on the concentration profile of the effluent from a chromatographic column, of increasing the amount applied, is illustrated in Fig. VI.2.

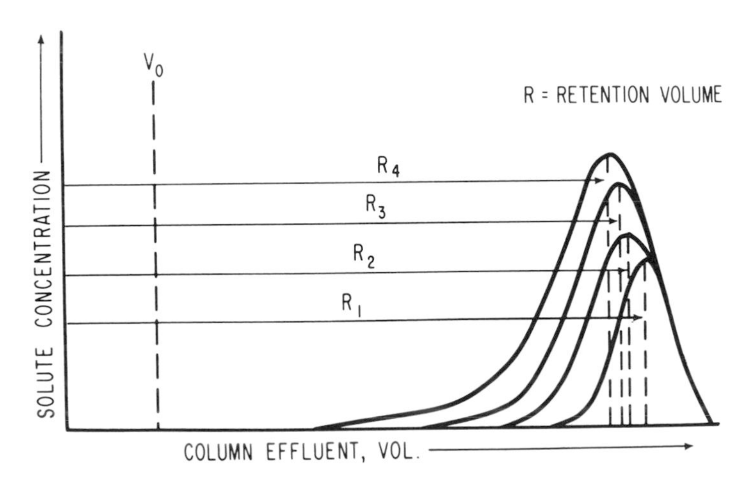

FIG. VI.2. Typical elution record. Effect of increasing column load.

The retention, R, or elution volume, V_e is the point of emergence of 50% of the solute. This is related to the distribution coefficient, G, between the stationary phase and the mobile phase by the expression

$$V_e = V_0 + V_s/G$$
$$V_s/G = V_e - V_0 = R_0$$

where V_s is the volume of the stationary phase and R_0 is the corrected elution volume or retention. Hence the elution volume corrected for void volume is directly proportional to G, the effective distribution coefficient.

A. LINEAR DISTRIBUTION COEFFICIENT. Since Eq. (VI.2) indicates that G is nonlinear with loading (θ) even if the adsorption sites are uniform, Snyder has proposed an arbitrary definition for a condition of uniform adsorption sites as the situation in which the percentage change in corrected elution volume (R_0) of two different sample weights divided by the logarithm of the sample weight ratio is less than 10%. Thus, for a tenfold increase in the weight of sample charged the ratio of corrected elution volumes should be greater than 0.9.

It can be seen that the uniformity of the adsorbent surface is reflected in the amount of solute which can be charged within the region of "linearity" as defined above. If a small portion of the surface has a higher affinity for the solute than the remaining surface, the region of linearity will include only that small portion.

The sequestering or inactivation of portions of the surface of silica or alumina by the addition of water has been studied extensively. For most systems the water is strongly bound to the surface and decreases the area available to other solvents or solutes. Thus a decrease in both total capacity and distribution coefficient is observed when water is added to these adsorbents. However, the remaining surface may be more uniform. This has been demonstrated for both silica and alumina by Snyder.

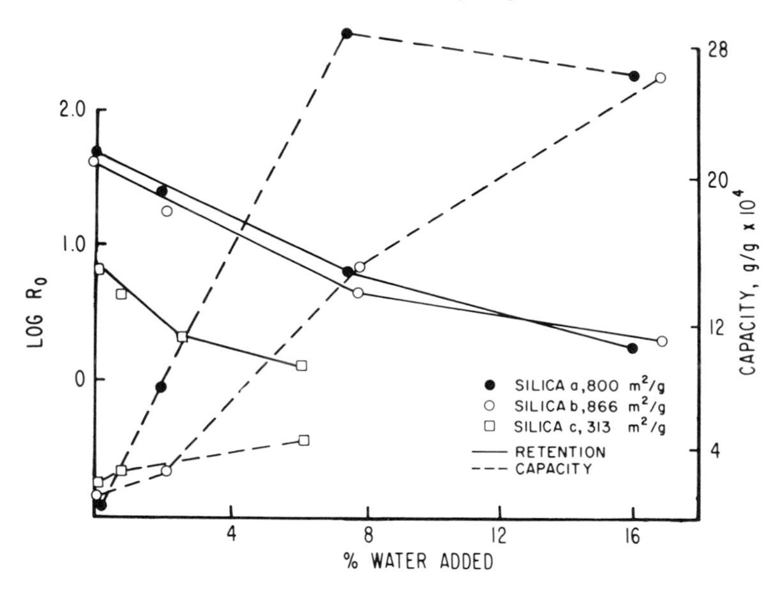

FIG. VI.3. Typical water deactivation. Retention and linear capacity as a function of added water in the system: silica, naphthalene–pentane.

An illustration of this effect with the system naphthalene/pentane and three samples of silica gel is given in Fig. VI.3 (*3*). If it is assumed that the area occupied by the same weight of either naphthalene or water is the same, then 1 % water corresponds to the equivalent 100×10^{-4} g/g of naphthalene. Since the maximum linear capacity observed is only about 30×10^{-4} g/g it is apparent that only a small portion of the surface is utilized in the linear region.

For two of the silica gel samples both R_0 and linear capacity change less and are more nearly the same in the region of 12–16% water which corresponds to the usual thin-layer plate stored under atmospheric conditions.

For a particular solute, solvent, and adsorbent system, the following relationship has been proposed:

$$\log R_0 = \log V_a + \alpha f(S, E) \qquad \text{(VI.4)}$$

in which V_a is the volume of an adsorbed monolayer covering the "active" area of the adsorbent and $\alpha f(S, E)$ is a term correcting for α, "surface activity" of V_a for the solute, S, and the solvent, (eluant, E). The absolute value of V_a is influenced by the treatment of the adsorbent. The addition of water is commonly used to decrease V_a. The equation indicates that different samples of adsorbent may have different affinities (R_0), even if V_a is corrected to the same value. This relationship is important since it indicates that for the separation of two solutes, the ratio of R_0 values may vary at different absolute values of R_0.

For a particular solvent–adsorbent system, even though the degree of separation of two compounds may be different at two values of V_a, the order of elution of the two solutes remains the same. However, use of another solvent can cause a reversal in the order of two solutes.

B. SURFACE AREA. Most useful adsorbents have surface areas in excess of 100 m^2/g. This area requires about 35 mg of water for complete coverage with a monolayer. Surface area measurements are frequently carried out by degassing the adsorbent under high vacuum at elevated temperatures and

TABLE VI.1

SURFACE AREA OF SEVERAL ADSORBENTS

Adsorbent	Surface area (m^2/g)
Darco G–60	1300
Graphite	30
Carbon black	7–100
Bone char	120
Alumina	100–600
Bentonite	20
Diatomaceous earth	4
Fuller's earth	130
Kieselguhr	20
Silica gel	300–800
Titanium dioxide	8
Florisil	300

then adding a known quantity of indicator. One of the most common procedures is that devised by Brunauer, Emmett and Teller (BET method); nitrogen gas is the indicator.

The reported areas for some common adsorbents and related substances are given in Table VI.1.

c. Surface Groups. Unfortunately it is difficult to characterize the chemical nature of the surface and it is apparent that the actual chemical elements existing on the surface contribute appreciably to the adsorption process. Surface types have been classified broadly as polar or nonpolar. In general, polar adsorbents contain surface groups capable of hydrogen bonding with adsorbed materials and this is probably the single most important property of this type of adsorbent. Nonpolar adsorbents have no hydrogen bonding properties and London dispersion forces are the most important in the adsorption mechanism. It has been suggested that silica gel, a typical polar adsorbent, contains four main types of surface elements. These are

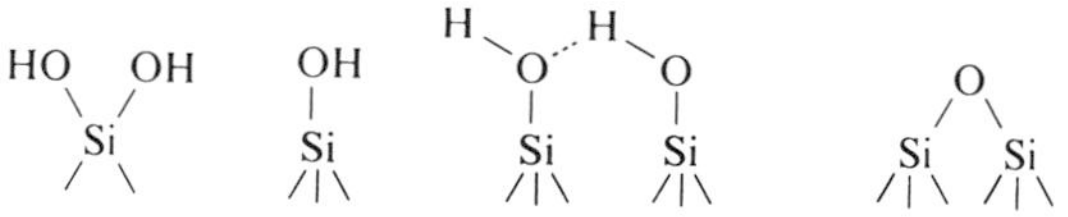

```
                              H       H
HO   OH       OH               \ O··· \ O              O
  \ /          |                 |      |             / \
   Si          Si                Si     Si           Si   Si
  /  \        /|\               /|\    /|\          /|\  /|\
```

Thus, the formation of hydrogen bonds between the surface and the adsorbed material contributes substantially to the adsorption process. Solvents capable of hydrogen bonding are good eluants. Some surface groups are weakly acidic and therefore can enter into ion exchange reactions at high pH. At high activation temperatures silica surfaces sinter and become less active. This may be due to a chemical change in the surface.

The alumina surface probably contains hydroxyl groups of similar structure. However, ion exchange reactions are greater than those observed with silica. Calcined alumina is strongly basic and may be acid-washed to produce a neutral or slightly acidic surface. A common acid washing procedure is to add sulfuric acid to a dilute aqueous suspension of basic alumina until pH 4 is obtained. Chromatography of salts containing anions other than sulfate on this type of alumina usually yields product containing sulfate ion.

The adsorption energy for aliphatic compounds containing various functional groups on alumina has been determined and was observed to increase in the following order: –S–, –O–, –C=N, $-CO_2^-$, –CO–, –OH, –N =, NH_2, –SO, $-CONH_2$ (*4*).

The reactivity of basic alumina was used by Klein and Knight (*5*) to prepare tritium-labeled keto steroids. In this case the alumina was treated with tritium oxide and the keto steroid applied to a small column in a suitable nonpolar solvent. Using cholest–7–en–3–one a high yield of tritium-labeled steroid was obtained.

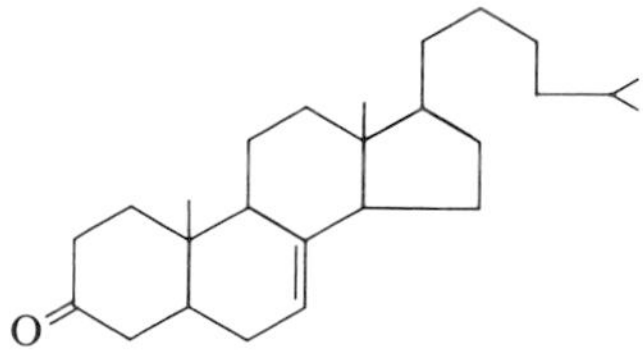

Cholest-7-en-3-one

Since the label is reduced substantially by alkaline treatment or rechromatography on unlabeled alumina, it is probably introduced by keto–enol tautomerism.

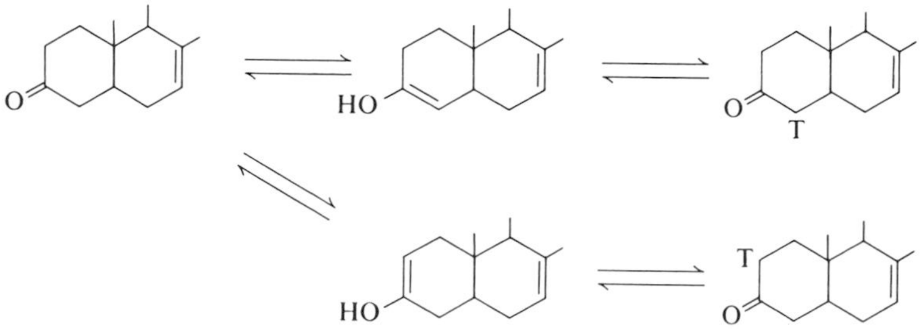

Although evidence that the enolization is important in the adsorption process was not obtained from these experiments, the fact that the exchange can take place indicates the possible participation of reactive surface groups in the adsorption process. In addition, a small percentage of label was found to persist after alkaline treatment of the keto steroid or was taken up by steroids which did not contain keto groups.

Chromatography of the acidic antibiotic, aureolic acid (*6*), on Florisil, a magnesium silicate adsorbent, yielded the magnesium salt, evidence that ion switching occurred.

Although most adsorbents are essentially inert inorganic substances, organic gels have been prepared and have limited usefulness as adsorbents. Some of the adsorption properties of ion exchange resins have been mentioned. In addition, phenol–formaldehyde polymers (Duolite S–30, Ionac D–100, and the like) have been used in sugar processing and the isolation of natural products. A neutral polystyrene (Amberlite XAD–2) and a lightly sulfonated polystyrene (Dow, NC–1266) are operable as nonpolar adsorbents. These materials adsorb organic compounds, especially neutral aromatics, from aqueous solution. The adsorbed materials are usually eluted by adding a water-miscible solvent.

Since these substances are available either in bead form or in sized granules, column operation is much more feasible than with activated carbon.

Some of the common adsorbents and their general utility are tabulated in Table VI.2.

TABLE VI.2

TYPES OF ADSORBENTS

Adsorbent	Type of solute preferred[a]	Useful solvents
Silica	Aromatic, polar, basic	Hydrocarbon, chlorinated hydrocarbon, alcohols
Alumina, basic	Aromatic, polar, acidic	Hydrocarbons, alcohols, water
Alumina, acid-washed	Aromatic, polar, basic	Hydrocarbon, alcohols, ketones, water
Carbon	Aromatic, polyunsaturated hydrocarbons, nonpolar to somewhat polar, acidic, or basic. Small, polar (especially neutral or amphoteric molecules) not adsorbed	Hydrocarbons, ketones, alcohols, water, pyridine
Magnesol, Florisil, etc.	Similar to alumina or silica; generally weaker adsorbents	Ketones, alcohol
Acidic clays, Fuller's earth, etc.	Basic, polar	Alcohols, water
Duolite S–30,[b]	Aromatic, neutral, basic	Water, alcohol, ketones
Ionac D–100	Aromatic, neutral, basic	
XAD–2,[b]	Aromatic, neutral, acidic	Water, alcohol, ketones
NC–1266[b]	Aromatic, neutral, acidic	Water, alcohol, ketones

[a] Molecular weight, generally less than 2000.

[b] Organic gels containing a relatively large proportion of aromatic residues; somewhat more specific than inorganic adsorbents. These are available from Diamond Shamrock, Permutit, Rohm and Haas, and Dow Chemical Co.

D. SURFACE TREATMENT. Covering the surface with water or some other strongly adsorbed material has already been mentioned for silica and alumina. Recently it has been shown that alumina which has been deactivated by the addition of water will generally revert to Brockmann Grade II (*25*) if anhydrous solvents are used. Thus a change in the adsorptivity of alumina could be obtained in the course of a chromatographic run.

The amount of water which must be added to a variety of solvents to maintain a particular alumina type is contained in Table VI.3 (*7*).

TABLE VI.3

WATER CONTENT (Vol. %) OF SOLVENTS IN EQUILIBRIUM WITH ALUMINA OF VARIOUS TYPES

Solvent	Alumina type				
	I	II	III	IV	V
Benzene, toluene	—	0.012	0.036	0.043	0.045
Chloroform	—	0.036	0.083	0.098	0.11
Methylene chloride	—	0.05	0.13	0.15	0.17
Ethyl ether	—	0.15	0.55	0.73	0.75
Ethyl acetate	—	0.57	2.06	2.4	2.6
Acetone	—	2.0	>20	—	—
n-Propanol	—	>20	—	—	—

The incorporation of other materials in the adsorbent can markedly influence utility and specificity of separation. The addition of silver nitrate to silica or Florisil improves the separation of unsaturated compounds. The addition of fatty acids to activated carbon was reported to improve the adsorption and elution of streptomycin (*8*). The incorporation of the acid–base indicator, methyl orange, in freshly precipitated silica was described by Martin and Synge (*26*). Kieselguhr containing boric acid has been used for the separation of sugars on TLC (*27*). Many other examples have been used in specific cases.

D. Substrate Properties

An attempt to apply generalities to all situations in which adsorbents may be used, although helpful, does not necessarily provide a solution for a specific separation problem. The ubiquitous use of thin-layer chromatography has provided a rapidly expanding bibliography of detailed examples of successful separations. However, the extensive use of polycomponent systems which usually produce a series of secondary fronts or gradient elution effects indicates that separation of a particular compound is likely to require careful selection of the adsorbent–solvent combination. In addition, since detection methods are not universal, many separations which are adequate for thin-layer chromatography are unsatisfactory for the preparation of pure compounds.

1. Effect of Molecular Size

If it is assumed that the adsorbate–solute complex consists of interacting areas of the surface with the adsorbed molecule, then the greater the number of possible interacting sites possessed by the solute, the greater the attraction and resulting adsorption. Thus, for a homologous series adsorption is influenced by molecular weight. For aliphatic compounds containing a single polar functional group (–COOH, –OH, $-NH_2$, –OH, –CHO, etc.), adsorption increases with nonpolar adsorbents such as activated carbon. On the other hand, if the number of active adsorption sites remains the same then relative adsorption decreases as the molecular weight increases. Hence, with silica and alumina, adsorption decreases with increasing molecular weight (or decreasing polarity) of simple compounds containing the same number and types of polar groups.

As a general rule the presence of aromatic or planar groups increases adsorption on both polar and nonpolar surfaces.

Since the adsorbed molecule must pentrate the adsorbent matrix, the adsorption of large molecules may be slight or, if appreciable, attain equilibrium very slowly. Separation, by adsorption processes, of materials having a molecular weight of >2000 is generally not desirable. Good separation of vitamin B_{12}, m. wt. 1400, is obtained on alumina chromatography and the material is readily adsorbed on activated carbon.

High molecular weight materials may be irreversibly adsorbed at low loading by activated carbon or organic gels.

2. *Steric Effects*

Steric effects are important in the separation of related compounds. Since the adsorption is a molecular property, isomeric compounds may be separated by adsorption methods, whereas other separation methods may fail. The separation of neomycin B and C using chromatography on carbon (*9*) or alumina (*10*) has been reported.

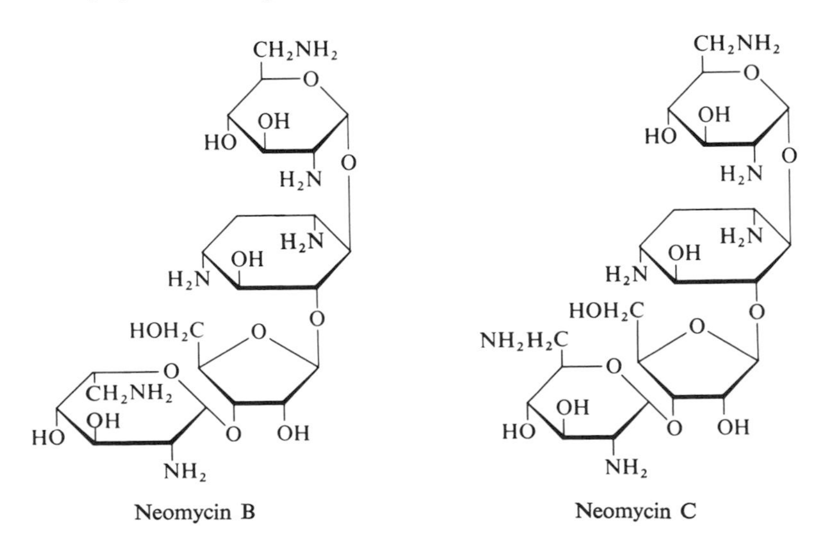

The same compounds are also separated by chromatography of the free bases on anion exchange resins. However, separation by conventional ion exchange chromatography is difficult since the pK of the basic groups in the two compounds is nearly identical and the interaction of the polar molecule with the resin backbone is limited.

As might be expected adsorption affinities of isomers may be reversed depending on the adsorbent (*11*, *12*). Thus, *trans*-azobenzene is more strongly adsorbed on activated carbon (methanol or water) than *cis*-azobenzene, whereas using alumina and petroleum ether the reverse order is obtained.

3. Martin's Relation

Although A. J. P. Martin originally considered that the distribution of a solute between two liquid phases and the expression for K, the distribution coefficient, were related to the free energy required to transfer 1 mole of the solute from one phase to the other, the fact that in adsorption processes one of the phases is solid does not invalidate the concept so long as the adsorbent surface is uniform, i.e., the free energy of transfer is constant. Since the solute is competing for surface sites with the solvent, the energy of solvent desorption must also be constant. The linear distribution concept introduced by Snyder can be seen as applying to that portion of the adsorbent surface which satisfies the above energy criteria or, more precisely, to that portion of the adsorption isotherm in which the net energy change is constant.

The total energy change includes factors of hydrogen bonding, London, or dispersion forces and other interactions, which will not be considered in detail. For any solute, the total energy, Δu_0, is a sum of the components. For a homologous series, $\Delta u_0 = \Delta u_1 + n\Delta u_2 + y\Delta u_3$, etc., in which Δu_1 could be due to the functional group and Δu_2 to repeating groups. Thus, analogous to liquid systems,

$$\log K = \Delta u_1 + nu\Delta_2 + \text{etc.} \tag{VI.5}$$

That this general relationship may apply to adsorption processes was indicated by Claesson (*13*), who studied the distribution coefficient for five different types of activated carbon and fatty acids, ethyl esters, dibasic acids and alcohols. For these compounds and four of the carbon samples the general relation

$$K = pq^n \tag{VI.6}$$

was proposed, in which p and q are constants; n is the number of carbon atoms in the compound; and K is the distribution coefficient of the Langmuir isotherm. Rewriting in the logarithmic form

$$\log K = \log p + n \log q \tag{VI.7}$$

a superficial resemblance to the Martin relationship is clearly indicated.

Brenner *et al.* (*14*) has discussed the application of Martin's relation to thin-layer chromatography. A constant $\Delta R_m(CH_2)$ was obtained with simple dinitrophenylamines using the system benzene–silica gel if the thin-layer plates were developed horizontally. However, tests with other series of compounds did not appear to yield constant ΔR_m values.

Further study indicated that the problems were in obtaining satisfactory chromatographic conditions rather than in the soundness of the theoretical approach. By limiting R_f values to those between 0.1 and 0.8, correcting the observed R_f value to account for the failure of the solvent near the front to completely saturate the adsorbent, and choosing solvent systems in which solvent demixing is not appreciable, constant $\Delta R_m(CH_2)$ values were obtained for a series of α-amino acids. Three systems—methanol, ethanol, or *n*-propanol and water (70:30) with silica gel G and with or without added acetic acid (up to 16%)—gave linear $\Delta R_m(CH_2)$ values. However, solvent demixing appeared important in system containing chloroform or in aqueous basic systems. Methods of correcting for the presence of secondary or tertiary solvent fronts have been suggested and appear useful for some systems.

Thus, it appears likely that the general Martin postulate applies to adsorption as well as solvent extraction if surface uniformity can be controlled and for compounds which do not have markedly different steric factors.

E. Solvent Properties

1. Relationship with Adsorbent and Substrates

The effect of changing the solvent in a three-component system in which no complex formation with the solute takes place is simply related to the affinity of the surface for the solvent. For group separation processes the general operating procedure is to carry out the adsorption step using a solvent more polar than the solute and a solvent less polar than the solute for the elution step (assuming that the solute is soluble in each case) for nonpolar adsorbents. For polar adsorbents the reverse procedure is used. For many natural products the orginal solution is aqueous, therefore the most successful adsorption steps are those carried out with a nonpolar adsorbent. Activated carbon is used for more extensively than other adsorbents for the treatment of aqueous solutions.

Dilute solutions of relatively polar compounds in less polar solvents may be treated with polar adsorbents for removal of the polar constituents.

For the elution step the solvent is ordinarily decreased in polarity for nonpolar adsorbents and increased in polarity for polar adsorbents. For the same solvent, the polarity of ionizable solutes may be changed by pH manipulation. Thus some adsorption–desorption steps combine both pH and solvent change to improve yields.

2. *Solvent-Adsorbent Interaction*

The heat of wetting has been used as an indication of the affinity of adsorbents for solvents. Data for some common solvents and adsorbents are given in Table VI.4.

TABLE VI.4

HEAT OF WETTING AND DIELECTRIC CONSTANT FOR POLAR ADSORBENTS

Solvent	Dielectric constant	Heat of wetting			
		Clay	Floridin	Alumina	Silica
Water	81	15.3	12.6	30.2	—
Methanol	31	15.3	11.0	21.8	27.6
Ethanol	25.8	14.7	10.8	17.2	24.5
Propanol	22.2	13.5	10.2	—	—
Acetone	21.5	13.5	8.0	27.3	—
Ethyl acetate	6.1	—	—	18.5	—
Diethyl ether	4.4	8.4	5.8	10.5	—
Chloroform	5.2	8.0	9.0	8.4	15.7
Benzene	2.3	8.1	5.8	4.6	10.8
Carbon tetrachloride	2.2	8.1	1.8	4.6	9.9
Petroleum ether	1.89	3.1	1.2	3.9	7.2

For polar adsorbents there is a good correlation between dielectric constant, an indication of polarity, and the heat of wetting. Exceptions are noted though and these may account for specificity in certain cases.

Eluotropic series have been proposed by various authors and are frequently used in a practical way in chromatography. Such series usually follow the heats of wetting as outlined above for polar adsorbents. For activated carbon the series in Table VI.5 may be used in the order of increasing elution ability.

TABLE VI.5

ELUOTROPIC SERIES FOR ACTIVATED CARBON

Water
Methanol
Ethanol
Propanol
Ethyl ether
Butanol
Ethyl acetate
n-Hexane
Benzene
Pyridine

Correlation of dielectric constant with eluting power has been questioned on the basis of observed exceptions. The data may be rationalized, however, by recognizing that in surface interactions molecular geometry is also important. Planar molecules entering into an adsorption reaction are more likely to have multiple site interaction and hence stronger adsorption than nonplanar molecules. Similarly, ring compounds, even though nonplanar, are apt to adsorb more strongly than would be predicted on the basis of dielectric constant. Displacement of such a molecule from the surface requires simultaneous loosening of all interacting sites.

Using evaporated metal film catalysis for deuterium exchange, Anderson and Kemball (*15*) have shown that with cyclohexane the most abundant partially deuterated product contains six deuterium atoms and very little monodeuterated product is obtained.

Practical experience has indicated that pyridine is an extremely effective eluant for both charcoal and alumina. That procedures indicated by the generalities outlined above are frequently practical is indicated by the following example, which is of historical interest.

The orginal isolation and purification procedure for the antibiotic, streptomycin, from fermentation broth of low potency contained only adsorption–desorption steps. The polar, basic antibiotic was adsorbed in high yield from aqueous solutions at pH 6–8 with activated carbon. The carbon adsorbate was then eluted with aqueous solvent mixtures, such as methanol or acetone, 30–50% at lower pH (2–3). The hydrochloride salt in methanol was then purified by chromatography in acid-washed alumina. This procedure separates streptomcyin A, a trisaccharide from streptomycin B, a tetrasaccharide.

Streptomycin B is more strongly retained during chromatography and better yields are obtained if water is added to the methanol developing solution after streptomycin A has been eluted.

Thus on the nonpolar adsorbent, adsorption is from a solvent more polar than the antibiotic and elution is effected by adding a less polar solvent to the mixture. For the polar adsorbent, adsorption is from a less polar solvent and elution is accomplished by increasing the polarity.

It should be noted that if the substrate is not adsorbed from water by activated carbon, adding a solvent is not likely to improve the adsorption. However, decreasing the polarity of the solute by pH manipulation can be of benefit.

Although the use of an eluotropic series and the application of the above generalities can be quite useful in many separation problems, the literature abounds with exceptions. These are of such magnitude that the validity of the general concept may be questioned. However, the exceptions can best be considered as indications of the complexity of the system. Certainly the surface area is of such diversity that specific effects can take place on a part of the surface. The magnitude of the solute–solvent interaction can be markedly different in certain cases. In addition, many apparent exceptions arise from chromatographic studies using mixed solvents in which the role of the second solvent may be much different than anticipated and the actual composition of the solvent in the adsorbent bed is unknown.

F. Quantitative Treatment of Adsorption Processes

The discussion above has pointed out several of the commonly accepted principles of adsorption processes. However, recently much progress has been made in quantification of the parameters of surface area, adsorbent activity, and solvent-eluting ability, particularly with silica and alumina adsorbents.

The concept of linear elution adsorbent chromatography has been briefly mentioned. Although many separation procedures are operated under conditions for which the distribution is far from linear, other, more difficult separations may require careful attention to details of surface activity and solvent selection so as to achieve linear conditions.

The quantitative aspects are much more important for fractionation than for group separation and the following summary is of interest primarily in the separation of compounds having similar physical and chemical properties. For the isolation of natural products some type of group separation is usually performed at an early stage which produces a concentrated mixture of compounds having similar chemical and physical properties. Adsorption chromatography can be applied most readily to substances which are somewhat solvent seeking. Since ion exchange chromatography is the best method for most highly polar substances, ion exchange and adsorption procedures are usually complementary and not competitive. Elution mixtures of changing composition, either gradient or stepwise, have frequently been employed in adsorption chromatography. For the separation of a single pure component a constant composition eluant is advisable and conditions tending to produce a linear distribution are desirable.

1. Sample-Adsorbent Weight Ratio

For monolayer coverage the total surface can adsorb about 0.3 mg/m^2 and an adsorbent with 400 m^2/g can adsorb up to one-eighth of its weight. Under the ideal conditions of the Langmuir isotherm, the distribution coefficient is nearly linear for the first 10% of available surface coverage (see Fig. VI.4), for a particular substance in the presence of an adsorbing solvent. If additional solutes are present, each has its own isotherm and in effect diminishes the area available to the desired solute. Assuming that the material being fractionated is 10% pure and that all of the impurities are adsorbed under linear conditions, a maximum ratio of impure solute to adsorbent of 1 to 80 could be used.

For chromatography a further assumption that the charged zone should not occupy more than 10% of the column length indicates that the ratio of 1 to 800 is desirable. Under these circumstances large columns are required and the product is necessarily obtained from the column in dilute solution.

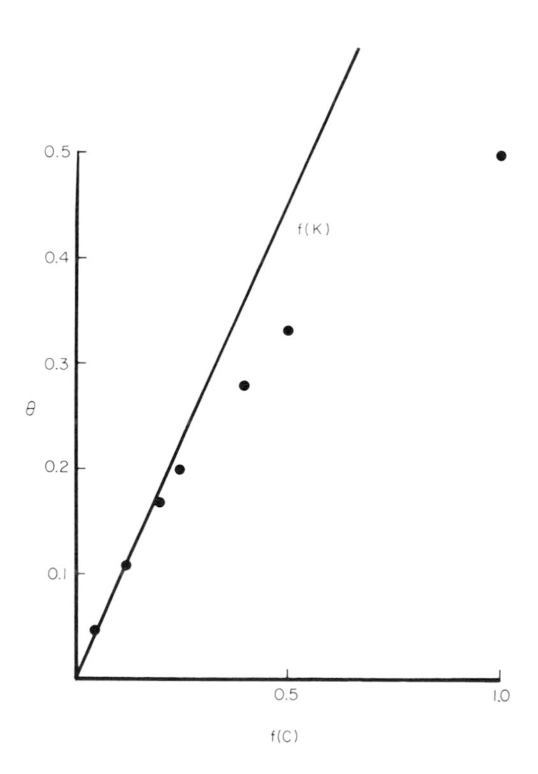

FIG. VI.4. Langmuir isotherm. Effect of load and distribution coefficient on linearity of *K*. Arbitrary units of *K* and *C*.

However, many satisfactory chromatographic separations have been obtained under conditions in which the distribution is far from linear and with solute–adsorbent ratios as low as 1 to 50. A study of the details of any adsorption chromatographic process can yield information on the separation obtained with different solvents and adsorbents, and may lead to a substantial increase in the weight of solute which can be charged to the chromatography.

It is interesting to consider the ratio of solute to adsorbent in thin-layer chromatography. Usually the adsorbent contains a large quantity of water. A

250 μ thick layer is commonly used and a zone width of 3–5 mm is obtained. For 7.5 cm travel ($R_f = 0.5$) this is about 50 mg of adsorbent.

The usual quantity of a single component used in a thin-layer chromatogram is 1–10 μg but quantities as high as 100 μg have been applied. This corresponds to a minimum adsorbent/solute ratio of 500 to 1. Almost all thin-layer systems can be made to "streak" by overloading. Progressively increasing the amount loaded produces a zone lengthening as the linear capacity of the adsorbent is exceeded. This does not preclude separation of two compounds in cases in which only one component is overloaded. If the overloaded zone is the most mobile it will move ahead of the second zone and separation will be satisfactory. However, if the overloaded zone is of lower mobility decreased retention may allow it to invade the more mobile zone and separation will be incomplete and become progressively less satisfactory as the load is increased.

2. *Overall Distribution Coefficient*

As with other chromatographic systems, a retention of about 10 column void volumes is most desirable for difficult separations since there is little increase in resolution by further increase in retention. As indicated previously, the retention is a function of adsorbent monolayer volume and activity, and solute and eluant properties

$$\log K = \log V_a + \alpha(\mathrm{S}, \mathrm{E}) \qquad \text{(VI.8)}$$

For a particular adsorbent and solute, the variables of adsorbent activity and solvent can be manipulated to achieve the desired retention. If the compounds being separated do not form a specific site adsorbate, the retention of the desired compound as well as the materials being separated will be influenced.

A. EFFECT ON SEPARATION FACTOR OF DEACTIVATING ADSORBENT OR INCREASING SOLVENT ELUTING POWER. Based on the assumption that the compounds being separated are similar in polarity and would not normally contain ionic groups, the effect on the separation factor, β, can be calculated using Eq. (VI.8). Sample calculations for two related compounds of similar size, one less retained and one more retained, are plotted in Fig. VI.5. In this figure the effects of inactivation of the adsorbent (added water) and solvent eluting power are compared. Values for log V_a and α according to Snyder (*16*) have been used.

The fact that the separation factor diminishes as K diminishes is not too surprising. However, it is seen that although slight differences in β are obtained depending upon the method chosen for regulating the distribution coefficient, these differences are negligible. Therefore, in ordinary practice the most convenient method may be used provided adsorbent deactivation does not

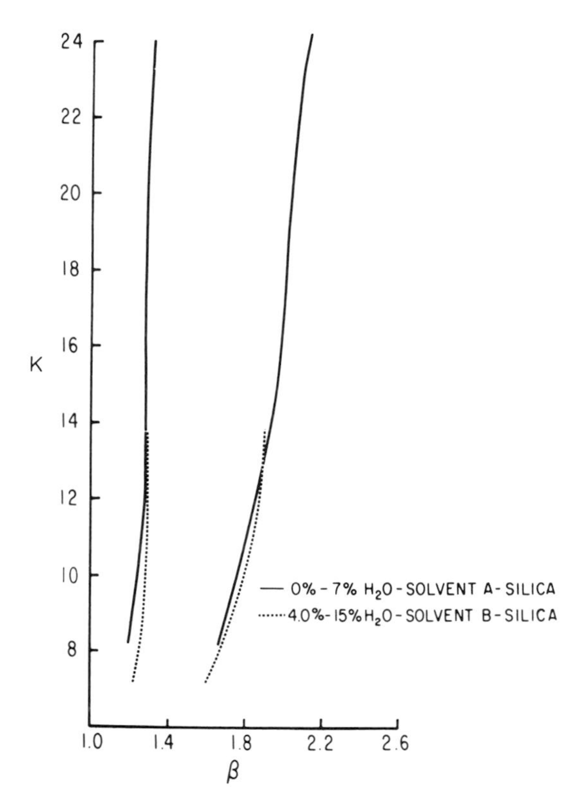

FIG. VI.5. Variation of β with adsorbent inactivation and solvent elution ability; log $K = V_a + \alpha f(S, E)$; V_a, α, and eluant variable.

proceed to the point that capacity is seriously diminished. If the available area is reduced to 100 m^2/g or less (75–95% inactivation for silica and 60–75% for alumina) further decreases in K should be achieved by choosing a more powerful eluant.

B. EFFECT OF SOLUTE SIZE. Although β values are not markedly affected by solvent changes if the compounds being adsorbed are similar in size, a large difference in β may be obtained with two compounds which give different adsorbed areas on the surface. Since most molecules are replaced on the surface by more than one solvent molecule and since, in effect, the surface is always covered by either solute or solvent, a mass action expression for the displacement of an adsorbed molecule is exponential.

Considering a unit of surface area A as that occupied by one molecule of solvent, X, then the surface reaction for the displacement of an adsorbed molecule Y, which occupies the area A, may be written as follows:

$$A_nY + n\,X \rightleftharpoons Y + n\,(AX) \tag{VI.9}$$

If it is assumed that an adsorbed monolayer of either solute or solvent has an equal thickness, the equilibrium constant for the surface reaction can be expressed as

$$K_a = \frac{[Y][AX]^n}{[A_nY][X]^n} \tag{VI.10}$$

For the solvent X and a second solute having larger area, n is larger and the equilibrium constant correspondingly increased. This represents a completely ideal situation in which interactions in the solvent phase are relatively unimportant and in which the surface sites are uniform for both solute and solvent. In addition, n is not likely to be an integer and a completely random array on the surface leads to areas that are inaccessible (without overlapping) to both solute and solvent.

For two solvents of exactly equal affinity for the surface but of different size, the larger solvent would more effectively displace any solute. If the solutes are also different in size the β value will be influenced with the solvent change. Since in many isolation problems the impurities may not have the same adsorbed size, it may be advantageous to change solvent size without influencing the distribution of the desired compound in cases in which increased resolution is required.

Since as solvent size increases fewer solvent molecules are required to displace a solute, as a general rule it is desirable to use small solvents for preliminary experiments since this is most likely to give the largest β between solutes of different size.

Approximately equal steps in diminishing K (for silica or alumina) are obtained with the series: pentane, benzene, isopropyl choloride, ethyl ether, chloroform, methylene chloride; acetone, methyl or ethyl acetate, dioxane, acetonitrile; isopropyl alcohol; ethanol; methanol. An eluotropic series for activated carbon has previously been mentioned.

II. APPLICATION OF ADSORPTION PROCEDURES

A. Group Separation

Adsorption procedures may be either of the group or fractionation type. Since there are literally thousands of different applications of adsorbents in the processing of natural products, the following examples have been selected as being illustrative of successful applications in cases in which other methods are unsatisfactory. Group separation by means of batch adsorption and elution steps usually cannot be operated to give high yields (90%) but good yields (60–80%) are frequently attainable. Although the optimization of a batch adsorption–elution process is complex, the Freundlich isotherm expression can usually be used to define both adsorption and elution conditions. Purification factors of 10–100 are frequently obtained. Since the group separation methods of solvent extraction and ion exchange usually have higher efficiency, adsorption is most frequently used when these processes are poor or when excessive quantities of solvent would be required.

1. Recovery of Vitamin B_{12}

Vitamin B_{12} is a neutral, rather polar substance. It is not extracted from water by the usual polar solvents but can be removed from water solution with either phenol or benzyl alcohol. It is prepared commercially by fermenttion. The usual finished fermentation mixture contains a few micrograms per milliliter of vitamin B_{12} equivalent. Therefore, for solvent extraction procedures, large quantities of solvent are required to concentrate a small amount of the vitamin. Immediately on completion of the fermentation the vitamin may be present in several different forms. One of the initial procedures used for the recovery of a concentrate from such fermentation liquor consisted of adsorption on activated carbon and elution with aqueous pyridine–acetic acid mixtures.

It was subsequently observed that good yields of the vitamin from activated carbon adsorbates could be obtained using a two-phase elution mixture. In this procedure the wet carbon was slurried with an *n*-butanol–water mixture and filtered. The *n*-butanol–water filtrate was separated and the water phase containing vitamin B_{12} retained. In this way the volume of eluate was reduced and impurities were discarded in the *n*-butanol, which was distilled and recycled. A detailed description of this procedure follows:

The filtered broth (*17*) was treated 2.5 g of activated carbon per liter (Darco G–60 or similar type) and stirred for 30 min. After adding an equal weight of filter aid (Hyflo Supercel), the rich carbon was filtered on a filter coated with filter aid. The carbon cake was thoroughly washed with water, removed from the filter, and slurried for 30 min with *n*-butanol using 5 ml/g of carbon. The slurry was filtered and the carbon cake reslurried with an additional 5 ml of butanol and 1 ml of water per gram. The filtrates are combined and the water layer which contains the vitamin separated and used in further processing steps, which usually consist of solvent extraction and chromatography.

2. Nucleic Acid Separation

Generalities concerning the use of adsorbents are usually subject to exceptions. The separation of ribonucleic acids (RNA) from deoxyribonucleic acids (DNA) by the use of activated carbon is an example which appears to contradict most of the generalities stated above. The molecular weight of the materials being adsorbed is about 25,000 and thus is much higher than the

2000 value previously mentioned. In addition, the presence of an additional hydroxyl group in RNA as compared with DNA would indicate that the latter should be preferentially adsorbed.

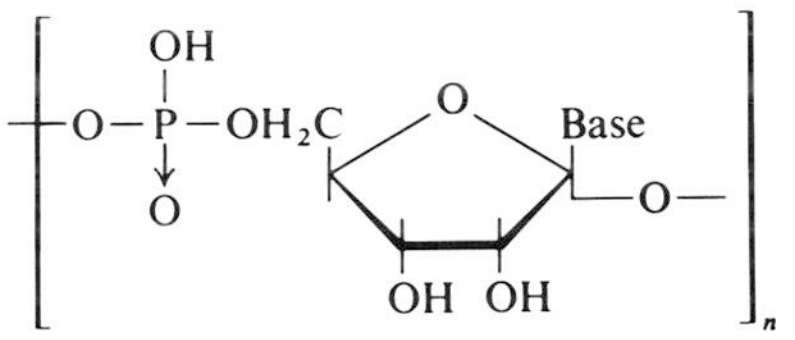

Ribonucleic acid (RNA)

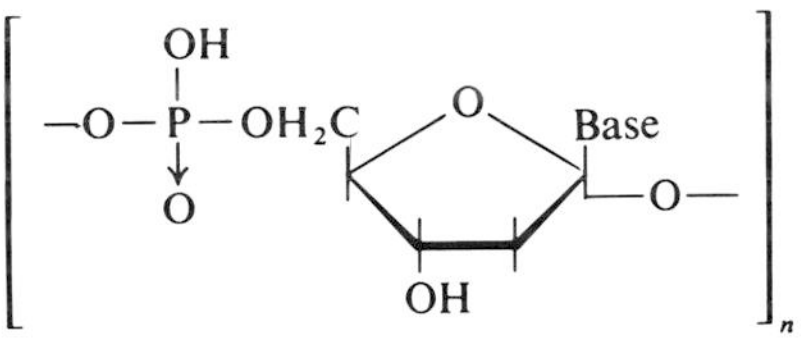

Deoxyribonucleic acid (DNA)

In fact RNA can be nearly completely removed from aqueous solutions of DNA with activated carbon (*18*). Investigations revealed that several types of activated charcoal can be used in a weight ratio of about 40–60 times the nucleic acid removed. At low ionic strength and neutral pH neither RNA nor DNA is adsorbed. As the ionic strength is increased, adsorption of both RNA and DNA increases but nearly maximum adsorption of RNA is obtained at 0.1–0.15 *M* sodium chloride, whereas DNA is poorly adsorbed in even 2 *M* sodium chloride. Recent investigation has shown that with very high ratios of adsorbent (400–600 to 1) most of the DNA also adsorbs even in 0.15 *M* sodium chloride (*19*). The adsorption of heated or sonicated DNA appears slightly greater than native DNA and the inherent viscosity of the unadsorbed DNA is greater than the adsorbed. Thus, the separation appears based, in part, on the molecular weight difference of the two materials, although detailed studies on the molecular weight of unadsorbed RNA (usually 1–5%) have not been made.

The adsorbed RNA can be recovered in good yield by elution of the carbon with 15% phenol (*20*).

For the separation a mixture of nucleic acids (1 mg/ml in 0.1–0.15 *M* sodium chloride) is treated with $\frac{1}{15}$–$\frac{1}{20}$ volume of washed activated carbon (Norite A, Norite SX25, or Darco G–60) and shaken for 1 hr at 2°C. The suspension is centrifuged. DNA is recovered from the supernatant. The RNA can be recovered by washing the carbon residue with water and then shaking at 2°C for 8–10 hr in 15 volumes of 15% phenol at pH 7–7.5. RNA can be recovered from the supernatant by adjusting to 1 *M* sodium chloride and precipitating with ethanol (3 volumes).

B. Fractionation Procedures

Historically, chromatography with adsorbents was first used in the separation of pigments of natural origin. Some of the first studies were with carotenes and polar oxide adsorbents. Since the carotenes are essentially hydrocarbons, the process was carried out in a hydrocarbon solvent and the eluting power of the eluant increased by adding increasing quantities of either benzene or ethanol. Observation of the zone movement of the colored compounds as the solvent composition was changed provided a ready method of controlling the procedure. However, for colorless compounds which cannot be observed on the column the mobility can only be determined after the material has been eluted and control of the process is much more difficult.

Operating procedures have evolved which generally use solvents of increasing eluting ability as defined above. This procedure has been applied to the separation of many types of organic compounds with vastly different chemical and physical properties. Many of the analytical chromatographic methods described in the literature have been superseded by gas–liquid chromatography which is usually much faster and more precise.

Examples of the actual preparation of pure substances by adsorption chromatography are relatively rare. Often chromatography is used to produce material of sufficient purity for crystallization. In such cases it is an essential step in the isolation even though pure product is not obtained directly. Usually high solute–adsorbent ratios are employed and large fractions collected.

1. Alkaloid Fractionation

The isolation of reserpine (*21*) is typical. The root extract was first purified by water and petroleum ether extraction. The insoluble residue was partitioned in a preparative countercurrent extraction using a chloroform–methanol–water system in six extraction vessels. By combining the two most polar of the twelve fractions a fivefold purification was achieved. The product was dissolved in benzene and chromatographed on alumina (Brockmann grade III, see Table VI.6) using a solute–adsorbent ratio of 1:23. The material was applied to the column in benzene–petroleum ether (4:1) solution. Continued washing of the column with the same solvent eluted inert oil and a small amount (about 1.5% of the charged solid) of a related substance, reserpinin, which was purified by crystallization. The eluates obtained with benzene and

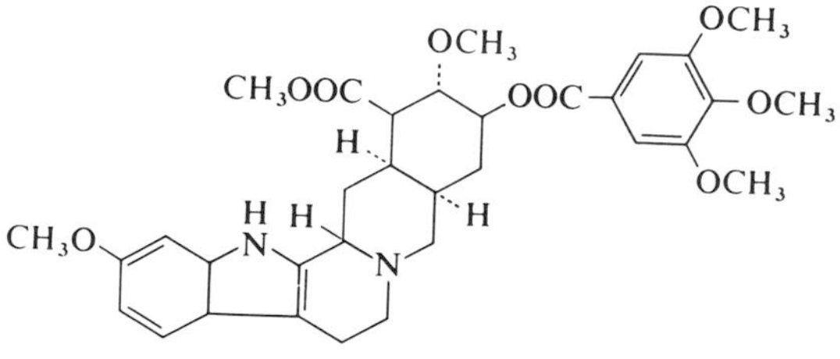

Reserpine

benzene–acetone (2:1) were combined after assay and yielded about 14% of the weight of the material charged to the chromatography as a crystalline product.

2. *Steroid Fractionation*

Although many procedures can be used to separate sterols, the usual isolation methods employing chromatography do not yield "pure" materials. The fractionation of sterol acetates on silica gel has been studied by Klein and Szczepanik (*22*), who observed that different samples of silica gel equilibrated at equal humidity gave different retention values in a benzene–hexane (1:5) system. A total of 13 sterols were studied with silica gel samples of different pore size. With large pore size, lowered retention and resolution were observed. Using 100-cm columns separation of two sterols could be accomplished if the ratio of retention volumes was greater than 1.20 and a low solute–adsorbent ratio was employed.

Decomposition of certain steroids was observed. Thus only about a 15% recovery of 7-dehydrocholesterol was obtained from silica chromatography. The probable enolization of sterols on basic alumina was mentioned earlier. However, the decomposition obtained on silica gel may be due to acid catalysis.

The separation of the acetate of 24, 25-dihydrolanosterol from lanosterol acetate was achieved

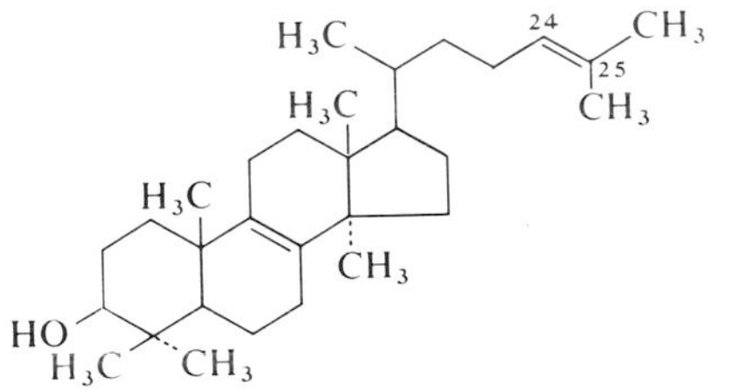

Lanosterol acetate

by using narrow pore silica (Davison #12) in benzene–hexane (1:5). Forty milligrams of the mixture was charged to a column (0.9 × 50 cm), containing about 20 g of adsorbent (500 to 1 ratio). The dihydro compound emerged from the column after 645 ml (about 20 column volumes) and was complete at about 975 ml. The more polar lanosterol appeared at 1125 ml (about 35 column volumes) and was complete at 1650 ml. These data indicate that additional crude mixture and lower retention volumes could be used to achieve the separation of these materials.

3. Dry Column Chromatography

The recent extensive use of thin-layer chromatography as a laboratory tool in following synthetic reactions has led to increased interest in preparative procedures using the same adsorbents and solvents. Several simple procedures, such as using thicker layers of adsorbent, a procedure which has proven difficult to scale up, or using wide plates with mechanical streaking or loading devices, have been proposed. In all of these procedures the preparation of appreciable quantities of material is difficult because of the large amount of plate area required.

A scale-up procedure which accommodates gram quantities of material has been proposed by Loev and Goodman (*23*). Columns are packed with dry adsorbents and developed as the thin-layer plate to the point at which all of the adsorbent has been wetted but no liquid has emerged from the column. The developed column may then be extruded and the desired zone eluted. A similar procedure using inverted columns has also been recommended (*24*).

A somewhat more convenient procedure is the use of a nylon tube for the chromatography. The thin-wall tubing transmits ultraviolet light and by incorporating a fluorescent mineral, such as zinc silicate, in the adsorbent, ultraviolet-absorbing materials may be detected. The column is then sectioned and the desired zone eluted.

The use of dry columns has the potential advantage that thin-layer chromatography can be used to find a solvent giving the desired separation.

TABLE VI.6

R_f of Indicator Dye with Samples of Alumina and Silica[a]

% of water added	Silica[b]	Alumina[c]	Brockmann Grade[d]	
			Silica	Alumina
0	0.15	0	I	I
3	0.22	0.12	—	II
6	0.33	0.24	—	III
8	—	0.46	—	IV
9	0.44	—	—	—
10	—	0.54	—	V
12	0.55	—	II	—
15	0.65	—	III	—

[a] Benzene solvent.
[b] Determined for *p*-dimethylaminoazobenzene.
[c] Determined for *p*-aminoazobenzene.
[d] Reference (*25*).

Thus many runs can be made using only small amounts of material in a relatively short time span for the evaluation of various solvent systems.

Since the finely divided adsorbent used for thin-layer plates is generally unsatisfactory for use in columns, a quick procedure for evaluating the adsorbent and correction to the same activity level is required.

The adsorbent is tested for activity by determining the R_f of a known compound using a capillary tube. A melting point tube is filled with adsorbent. The adsorbent in the open end is immobilized by wetting with benzene. The closed end is broken off, dipped in a benzene solution of dye, and then placed in benzene for development.

For silica, the dye used is *p*-dimethylaminoazobenzene and for alumina, *p*-aminoazobenzene is used. The same dyes are used for calibration of the thin-layer plates. Water is added to adjust the adsorbents to the same activity level as the plate. The amount of water needed for a particular sample of adsorbent can be determined from the trial run. Typical data for samples of alumina and silica are contained in Table VI.6.

The correlation of the above data to relation (VI.4) is indicated in Fig. VI.6, in which R_f has been converted to R_m. It is apparent that the R_m and corresponding retention for both adsorbents is influenced by the water content of the adsorbent. Since the surface areas of the two adsorbents are different the slopes of the lines plotted against water content are different. The activities of thin-layer plates handled in the laboratory are indicated in the figure.

Several workers have reported that different thin-layer mobilities are

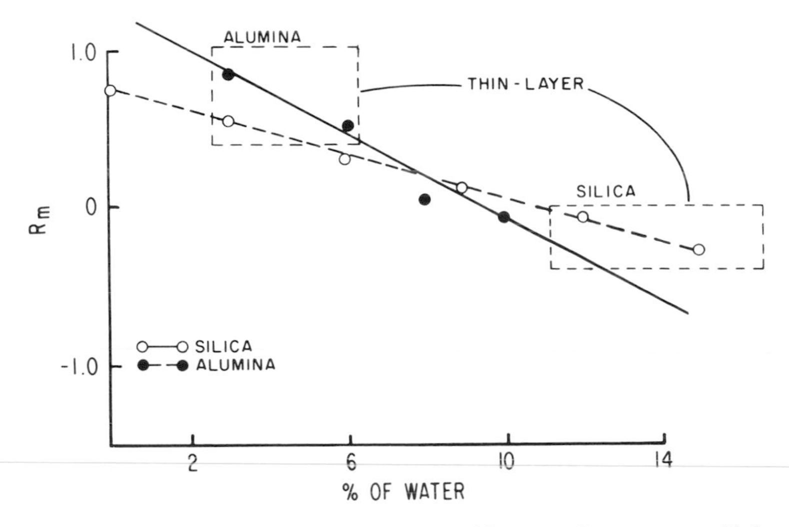

FIG. VI.6. Observed adsorbent inactivation with water; R_m vs. water added.

obtained when the sandwich type of apparatus is compared with the usual equilibrated tank. In the latter case the adsorbent surface is in contact with vapor of the developing solvent whereas vapor equilibration usually does not occur in the sandwich apparatus. The difference is greater with mixed solvents. Therefore, as a general rule mixed solvents should be avoided if possible in dry column procedures. If mixed solvents are used, it is likely that thin-layer plates developed in a sandwich type apparatus would correlate more exactly with column behavior than plates developed in an open tank. In addition differences in mobility due to solvent demixing effects may occur due to differences in the adsorbents used. In spite of these factors the dry column procedure remains an attractive method of separating unknown mixtures.

Loev and Goodman (*23*) have suggested the following general procedure:

Silica columns of 40 cm length are approximately equivalent in resolving power to columns of alumina of 20 cm length. The adsorbent is tested by the capillary method mentioned above, and the amount of water needed to increase the R_f to that with the range of thin-layer plates (see Fig. VI.6) is added to the adsorbent in a closed flask and the mixture equilibrated by shaking for 30 min. If the R_f difference of the compounds being separated were about 0.4, a column of about $\frac{1}{2}$-inch diameter would be used per gram of mixture to be separated. If the R_f difference were about 0.1, then a 1-inch diameter column would be used.

The adsorbent is loaded into the column with tamping. The material to be chromatographed is deposited on about 5 times its weight of adsorbent by making a slurry and evaporating the solvent. The dried solid is added mechanically to the top of the dry column.

The column is developed by carefully adding solvent to the top of the column. Development is complete in 45–90 min when the solvent front reaches the bottom of the column. The desired material is recovered by sectioning the column and eluting with a polar solvent such as methanol.

Although many additional examples could be provided, the general principles of the use of adsorption processes have been illustrated and some insight into the mechanism of the separation provided. Further details on adsorption separation methods may be obtained in the general reference texts.

REFERENCES

1. Langmuir, I., *J. Am. Chem. Soc.* **38**, 2221–2295 (1916); *J. Am. Chem. Soc.* **40**, 1361 (1918).
2. Snyder, L. R., *J. Chromatog.* **5**, 430–441 (1961); *J. Chromatog.* **6**, 22–52 (1961).
3. Snyder, L. R., *Separ. Sci.* **1**, 191–218 (1966).
4. Snyder, L. R., *J. Chromatog.* **23**, 388–402 (1966).
5. Klein, P. D., and Knight, J. C., *J. Am. Chem. Soc.* **87**, 2657–2661 (1965).

6. Philip, J. E., and Schenck, J. R., *Antibiot. Chemotherapy* **3**, 1218–1220 (1953).
7. Hesse, G., and Roscher, G., *Z. Anal. Chem.* **200**, 3–9 (1964).
8. Weiss, D. E., *Discussions Faraday Soc.* **7**, 142–151 (1949).
9. Ford, J. H., Bergy, M. E., Brooks, A. A., Garret, E. R., Alberti, J., Dyer, J. R., and Carter, H. E., *J. Am. Chem. Soc.* **77**, 5311–5314 (1955).
10. Dutcher, J. D., Hosansky, N., Donin, M. H., and Wintersteiner, O., *J. Am. Chem. Soc.* **73**, 1384–1385 (1951).
11. Freundlich, H., and Heller, W., *J. Am. Chem. Soc.* **61**, 2228–2230 (1939).
12. Zechmeister, L., Frehden, O., and Jorgensen, P. F., *Naturwissenschaften* **26**, 495 (1938).
13. Claesson, S., *Arkiv Kemi, Mineral. Geol.* **A23**, No. 1 (1946).
14. Brenner, M., Niederwieser, A., Pataki, G., and Weber, R., "Thin-Layer Chromatography" (E. Stahl, ed.), English Ed., pp. 75–133. Academic Press, New York, 1965.
15. Anderson, J. R., and Kemball, C., *Proc. Roy. Soc.* (*London*) **A226**, 472–489 (1954).
16. Snyder, L. R., "Principles of Adsorption Chromatography," Dekker, New York, 1968.
17. Wolf, F. J., U.S. Patent No. 2,530,416, (1950).
18. Zamenhof, S., and Chargaff, E., *Nature* **168**, 604–605 (1951).
19. Rosenkranz, H. S., and Rosenkranz, S., *J. Chromatog.* **30**, 549–555 (1967).
20. Dutta, S. K., Jones, A. S., and Stacey, M., *Biochim. Biophys. Acta* **10**, 613–622 (1953).
21. Dorfman, L., Furlenmeier, A., Huebner, C. F., Lucas, R., MacPhillamy, H. B., Mueller, J. M., Schlittler, E., Schwyzer, R., and St. Andre, A. F., *Helv. Chim. Acta* **37**, 59–75 (1954).
22. Klein, P. D., and Szczepanik, P. A., *J. Lipid Res.* **3**, 460–466 (1962).
23. Loev, B., and Goodman, M., *Chem. Ind.* (*London*) pp. 2026–2032 (1967).
24. Bhalla, U. K., Nayak, U. R., and Dev, S., *J. Chromatog.* **26**, 54–61 (1967).
25. Brockmann, H., and Schodder, H., *Chem. Ber.* **74**, 73–78 (1941).
26. Martin, A. J. P., and Synge, R. L. M., *Biochem. J.* **35**, 1358–1368 (1941).
27. Morris, L. J., *Chem. Ind.* (*London*) 1238 (1962).

Appendix I *Dielectric Constants of Some Common Solvents (at 25°C)*

		H Donor	H Acceptor
Hydrocyanic acid	114	+	+
Formamide	109	+	+
Water	78.5	+	+
Furfural	46	+	+
Glycol	37	+	+
Methanol	32.6	+	+
Ethanol	24.3	+	+
Acetone	20.7	−	+
n-Butanol	17.8	+	+
1-Hexanol	13.3	+	+
Pyridine	12.3	−	+
Methylamine	11.4	−	+
Phenol	9.8	+	+
Aniline	6.9	−	+
Acetic acid	6.2	+	+
Ethyl acetate	6.0	−	+
Chloroform	4.8	+	−
Propyl ether	3.3	−	+
Triethylamine	2.4	−	+
Benzene	2.284	−	−
Cyclohexene	2.2	−	−
Cyclohexane	2.0	−	−
n-Hexane	1.89	−	−

Appendix II *Ion Exchange Resins*[a]

CATION EXCHANGE RESINS

Moisture content (%)	% DVB	Screen mesh[b] (particle size, μ)	Trade name[c]
		I. *Strong Acid Type*	
a. Polystyrene, gel, $-SO_3H$			
85–92	1	20–60	Dowex 50WX1, BioRad AG-50WX1
75–85	2	20–50	Dowex 50WX2, BioRad AG-40WX2
		50–100	Dowex 50WX2, BioRad AG-50WX2
		80–200	Dowex 50WX2, BioRad AG-50WX2
65–75	4	20–50	Dowex 50X4, BioRad AG-50WX4
		40–80	Dowex 50X4, BioRad AG-50WX4
		60–140	Dowex 50X4, BioRad AG-50WX4
		100–200	Dowex 50X4, BioRad AG-50WX4
		200–400	Dowex 50X4, BioRad AG-50WX4
		(30–35 μ)	BioRad AG-50X4
		(20–30 μ)	BioRad AG-50X4
60–65	5	200–400	BioRad AG-50WX-5
55–60	6	16–50	Duolite C-25D, Amberlite XE-100
50–55	8	20–50	Dowex 50WX8, BioRad AG-50WX8, BioRex RG-50WX8,[d] Permutit Q
50–55	8	40–80	BioRad AG-50WX8
		60–140	Dowex 50WX8, BioRad AG-50WX8
		100–230	Dowex 50WX8, BioRad AG-50WX8
		230–400	BioRad AG-50WX8
45–50	10	20–50	Dowex 5X10, Dowex 50WX10, Amberlite IR-120, Amberlite IR-121, Amberlite IR-122

CATION EXCHANGE RESINS (*Continued*)

Moisture content (%)	% DVB	Screen mesh[b] (particle size, μ)	Trade name[c]
		100–500	BioRad AG-50WX10, Ionac C-240, Ionac C–249, Ionac C-253, Ionac C-257, Ionac Cl-295, Amberlite IR-120
		<325	Amberlite IR-120
40–45	12	20–50	BioRad AG-50WX12
		40–80	BioRad AG-50WX12
		60–140	Dowex 50WX12, BioRad AG-50WX12
		140–170	Dowex 50WX12, BioRad AG-50WX12
		230–400	BioRad AG-50WX12
		(25–32 μ)	BioRad Aminex 50WX12
		(21–29 μ)	BioRad Aminex 50WX12
35–40	16	20–50	Amberlite IR-124, Ionac C-250, Ionac C-258, Ionac C-255, Dowex 50WX16

Moisture content (%)	Screen mesh	Trade name
b. Polystyrene, macroreticular, $-SO_3H$		
46–50	16–50	Amberlite 200
		Amberlite 200C
		Amberlite 252
c. Sulfonated coal		
14–20	16–50	Ionac C-150
d. Cellulose or dextran, $-OC_2H_4SO_3H$		
89–90	120–400	SE-Sephadex C-25
96–98		SE-Sephadex C-50, Cellex SE
e. Phenolic, $-CH_2SO_3H$		
—	20–50	BioRex 40, Duolite C-1
—	50–100	BioRex 40
—	100–200	BioRex 40
—	200–400	BioRex 40
	II. *Intermediate Acid Strength*	
a. Polystyrene, phosphoric granular		
—	68–76	BioRex 63
b. Cellulose, phosphoric		
—	—	Cellex P

CATION EXCHANGE RESINS (*Continued*)

Polymer type	Particle type	Screen mesh	Trade Name
III. *Weak Acid Type*			
a. Methacrylic acid–divinylbenzene, –COOH			
p$K \sim 6.0$	Spherical	20–50	Amberlite IRC-50
	Granular	100–500	Amberlite IRP-64
	Granular	< 325	Amberlite IRP-64M
b. Acrylic acid–divinylbenzene, –COOH			
p$K \sim 5$	Spherical	20–50	Amberlite IRC–84, Amberlite XE-222, Duolite ES-80, BioRex 70
	Granular	50–100	BioRex 70
		100–200	BioRex 70
		200–400	BioRex 70
		< 400	BioRex 70
c. Miscellaneous weak acid resins			
Condensation	Granules	16–50	Ionac C–265
Dextran, $-OCH_2COOH$	Spherical	120–400	Sephadex C-25
		120–400	Sephadex C-50
Cellulose, $-OCH_2COOH$	Fibrous	—	Cellex CM, Cellulose CM-22, Cellulose CM-23
		200–400	Cellulose CM-32, Cellulose CM-52

ANION EXCHANGE RESINS

Cross-link (%)	Moisture (%)	Particle type	Size	Trade name
I. *Strong Base*				
a. Polystyrene—type I, $CH_2\overset{+}{N}(CH_3)_3$				
1	80–90	Spherical	20–60	BioRad AG-1X1, Dowex 1X1
2	70–80		20–60	BioRad AG-1X2, Dowex 1X2, Amberlite XE-238
			60–140	BioRad AG-1X2, Dowex 1X2
			80–200	BioRad AG-1X2, Dowex 1X2
			200–325	BioRad AG-1X1, Dowex 1X2

ANION EXCHANGE RESINS *(Continued)*

Cross-link (%)	Moisture (%)	Particle type	Size	Trade name
4	60–70		20–50	BioRad AG-1X4, Dowex 1X4, Amberlite IRA-900, Amberlite IRA-904, Duolite ES-111, Amberlite A-26 (macroreticular) Amberlite IRA-401S
4	60–70	Spherical	40–80	BioRad AG-1X4, Dowex 1X4
			60–140	BioRad AG-1X4, Dowex 1X4
			100–230	BioRad AG-1X4
			230–325	BioRad AG-1X4
—	50–60		16–20	BioRad AG-21K, Dowex 21K Amberlite IRA-401, Ionac A-540, Ionac A-544, Ionac A-548, Duolite A101D
			20–40	BioRad AG-21K, Dowex 21K
8	40–50	Spherical	16–50	BioRad AG-1X8, Dowex 1X8, Ionac A-546, Ionac A-000, Amberlite IRA-400, Amberlite IRA-400C
			40–80	BioRad AG-1X8, Dowex 1X8
			80–140	BioRad AG-1X8, Dowex 1X8
		—	140–325	BioRad AG-1X8, Dowex 1X8
		—	< 230	BioRad AG-1X8, Dowex 1X8
—	40–50	Granular	100–500	Amberlite IRP-67
			< 325	Amberlite IRP-67M
10	35–40	Spherical	40–80	BioRad AG-1X10
			60–140	BioRad AG-1X10
			140–325	BioRad AG-1X10
			< 230	BioRad AG-1X10

b. Polystyrene—type II, $CH_2-\overset{+}{N}(CH_3)_2(CH_2CH_2OH)$

Cross-link (%)	Moisture (%)	Particle type	Size	Trade name
4	55–60	Spherical	20–50	BioRad AG-2X4, Dowex 2X4, Amberlite IRA-901 (R)
	47–50		16–50	Ionac A-550, Ionac A-5
7	40–45		20–50	Amberlite IRA-410
8	34–40		20–50	BioRad AG-2X8
			40–80	BioRad AG-2X8
			60–140	BioRad AG-2X8
			140–235	BioRad AG-2X8
10	28–36	Spherical	40–80	BioRad AG-2X10
			60–140	BioRad AG-2X10
			140–235	BioRad AG-2X10

ANION EXCHANGE RESINS (*Continued*)

Cross-link (%)	Moisture (%)	Particle type	Size	Trade name
c. Polystyrene, $CH_2\overset{+}{N}$ (ring)				
—	46–54	Spherical	16–50	BioRex 9
		Granular	50–100	BioRex 9
			100–200	BioRex 9
			200–325	BioRex 9
—	—	Spherical	16–50	Ionac A-580
			10–20	Ionac A-590
d. Dextran				
—	—	Spherical	140–400	QAE Sephadex A-25
			140–400	QAE Sephadex A-50
II. *Weak and Intermediate Base Strength, Quaternary and Ternary*				
—	60–66	Spherical	16–50	Duolite A-57
—	50–60	Spherical	20–50	BioRex 5
		Granular	50–100	BioRex 5
			100–200	BioRex 5
			200–400	BioRex 5
—	—		16–50	Ionac A-300, Ionac A-302
III. *Weak Base*				
a. Polystyrene				
—	46–55	Spherical	20–50	Amberlite IRA-93 (R)
—	40–45	Spherical	20–50	Amberlite IRA-45, Amberlite IRA-21 (R)
—	25–35		16–40	BioRad AG-3X4, Dowex 3
			60–140	BioRad AG-3X4
			140–325	BioRad AG-3X4
b. Acrylic				
—	57–63	Spherical	20–50	Amberlite IRA-68, Amberlite XE-258
c. Phenolic–polyamine				
—	—	Granular	100–500	Amberlite IRP-58
			< 325	Amberlite IRP-58M
d. Condensation polymer				
—	—	Granular	16–50	Ionac A-260

ANION EXCHANGE RESINS (*Continued*)

Cross-link (%)	Moisture (%)	Particle type	Size	Trade name
e. Dextran and cellulose, DEAE–$OCH_2CH_2N(C_2H_5)_2$				
—	—	Spherical	160–400	DEAE Sephadex A-25
			140–400	DEAE Sephadex A-50
—	—	Fibrous	—	Cellex D, Cellulose DE-22, Cellulose DE-23
—	—	Spherical	100–200	Biogel DM-2
			100–200	Biogel DM-30, high cap.
			100–200	Biogel DM-30, low cap.
			100–200	Biogel DM-100, high cap.
			200–400	Cellulose DE-32, Cellulose DE-52

[a] Commercially available from United States' manufacturers.

[b] Screen size (U.S. standard mesh)

Screen size:	16	20	40	50	80	100	140	200	275	325	400
Particle size, μ:	1190	840	420	297	177	149	105	74	53	44	37

[c]

Resin trade names	Manufacturer
Dowex	Dow Chemical Co., Midland, Mich.
Amberlite	Rohm and Haas Co., Philadelphia, Pa.
Sephadex	Pharmacia Fine Chemicals, Inc., Piscataway, N.J.
Ionac	Ionac Chemical Corp., Birmingham, N.J.
Duolite	Diamond Shamrock Chemical Co., Redwood City, Calif.
Cellex and Cellulose DE and CM	Reeve Angel, Clifton, N.J.
BioRad	BioRad Labs., Richmond, Calif.
BioRex	BioRad Labs., Richmond, Calif.
Biogel	BioRad Labs., Richmond, Calif.

[d] Available in hydrogen, ammonium, lithium, or lithium-7 forms.

Appendix III *General Thin-Layer and Paper Chromatographic Techniques*

The standardization of chromatographic techniques is important to ensure reproducible results and to enable the characterization procedure to be adequately described. The following is a description of the most important variables which should be controlled.

The solvents should be carefully prepared and stored under stable conditions. Reactive solvents should be prepared fresh for each chromatographic run. The solvent composition, chromatography support, and approximate development time should be recorded. Abnormalities occurring at the time should be noted.

The amount of material applied to the chromatogram should be measured using a micropipette or similar device. It is usually desirable to include known standards in each run.

Spray reagents should be applied lightly and evenly to the chromatogram surface and development conditions standardized. Permanent records may be prepared by photographing the developed chromatogram, by tracing the chromatography with tracing paper, or by some other fixing technique.

The most universal visualization methods are mentioned below. Many more specific methods are available for use in determining the general class of compound or locating the desired zone in the presence of impurities.

1. Direct Visualization

View in a black box using ultraviolet light for the detection of both absorption and fluorescence. This may be enhanced by using light of different wavelengths and adsorbents containing phosphors. Exposure of the chromatogram to acidic (hydrochloric acid) or basic (ammonia) vapors may increase sensitivity.

After viewing the air-dried chromatogram, heating may reveal zones previously undetected.

2. Iodine Vapor

The dried chromatogram is placed in a closed tank containing iodine crystals. Compounds forming iodine complexes appear darker than the background during both iodine deposition in the tank and on iodine evaporation after removal from the tank.

3. Acid Charring

This method is useful only for thin-layer chromatography. The plate is sprayed with dilute sulfuric acid (1–2 *N*) or ammonium sulfate and heated at 120°C to develop charred zones.

4. *Potassium Permanganate*

Spray with 0.5% potassium permanganate. Transient zones may appear white on a pink background. After standing for 15 min a second spray of 0.2% bromphenol blue, which indicates acidic zones, is both more sensitive and stable than the primary zone.

5. *Fluorescence Quenching*

If a phosphor is not present in the chromatography support, fluorescence quenching may be observed by spraying the chromatogram with one of the following fluorescent substances:

(a) 0.5% solution of rhodamine B in ethanol

(b) 0.2% solution of 2′, 7′-dichlorofluorescein in ethanol

(c) 0.02% solution of 4-methylumbelliferone in 35% ethanol; intensify by exposure to ammonia vapor

6. *Ehrlich's Aldehyde (Nitrogen-Containing Hetrocycles)*

A 1% solution of *p*-dimethylaminobenzaldehyde in 50% ethanol–conc. hydrochloric acid is sprayed lightly on the chromatogram. After a few minutes at room temperature the chromatogram may be heated. Contrasting zones appear on a pinkish background. A variety of colors may be obtained.

The reagent is unstable at room temperature and should be freshly prepared or stored at −30°C.

7. *2,4-Dinitrophenylhydrazine (Aldehydes and Ketones)*

A solution (0.2–0.4%) of 2,4-dinitrophenylhydrazine in 2 *N* hydrochloric acid is sprayed on the chromatogram. May be heated to obtain yellow zones of aldehydes and ketones.

8. *Modified Rydon-Smith Test for Amides and Peptides*

A 0.1% solution of *tert*-butylhypochlorite in cyclohexane is sprayed on the chromatogram. After heating for a few minutes at 95°–100°C, the chromatogram is sprayed with a solution containing 0.2% potassium iodide and 0.1% starch. Blue zones appear on a white background.

9. *Phosphotungstic Acid (Lipids)*

After spraying with a solution of phosphotungstic acid in ethanol (5–20%) the chromatogram is heated at 70°C.

10. *Aniline Phthalate (Reducing Sugars)*

After spraying with a solution of 0.9 g of aniline and 1.6 g of phthalic acid in 100 ml of water saturated with butanol the chromatogram is heated at 100°–105°C for 10 min.

A modification of the method utilizes an equimolar quantity of *p*-anisidine in place of aniline.

Sugars appear as gray to grayish blue. Colors are somewhat characteristic.

11. *Periodate Reaction (cis-Glycols, etc.)*

After spraying with a solution of sodium metaperiodate (0.1%) and drying, the chromatogram is sprayed lightly with benzidine reagent. White zones appear on a blue background.

The benzidine reagent is freshly prepared by dissolving 1 g of benzidine in 30 ml of ethanol and diluting with 25 ml of water, 11 ml of acetone, and 0.5 ml of 1 *N* hydrochloric acid.

Appendix IV *Conversion Table for R_f and R_m*

R_f	0.01	0.02	0.03	0.04	0.05	0.06	0.07	0.08	0.09	0.10	
0.01–0.10	2.00	1.69	1.51	1.38	1.28	1.20	1.12	1.06	1.01	0.95	
0.11–0.20	0.91	0.87	0.83	0.79	0.75	0.72	0.69	0.66	0.63	0.60	
0.21–0.30	0.58	0.55	0.53	0.50	0.48	0.45	0.43	0.41	0.39	0.37	
0.31–0.40	0.35	0.33	0.31	0.29	0.27	0.25	0.23	0.21	0.19	0.18	
0.41–0.50	0.16	0.14	0.12	0.11	0.09	0.07	0.05	0.04	0.02	0.00	R_m
0.51–0.60	−0.02	−0.04	−0.05	−0.07	−0.09	−0.11	−0.12	−0.14	−0.16	−0.18	
0.61–0.70	−0.19	−0.21	−0.23	−0.25	−0.27	−0.29	−0.31	−0.33	−0.35	−0.37	
0.71–0.80	−0.39	−0.41	−0.43	−0.45	−0.48	−0.50	−0.53	−0.55	−0.58	−0.60	
0.81–0.90	−0.63	−0.66	−0.69	−0.72	−0.75	−0.78	−0.83	−0.87	−0.91	−0.95	
0.91–1.00	−1.01	−1.06	−1.12	−1.20	−1.28	−1.38	−1.51	−1.69	−2.00	—	

Author Index

Numbers in parentheses are reference numbers and indicate that an author's work is referred to, although his name is not cited in the text. Numbers in italics show the page on which the complete reference is listed.

C

D

E

F

G

M

N

O

P

Subject Index

H

I

R

S

T

U

V